Fracture Kinetics of Crack Growth

MECHANICAL BEHAVIOR OF MATERIALS

VOLUME 1

Managing Editor:
A. S. Krausz, University of Ottawa, Ontario, Canada

Advisory Board:
M. F. Ashby, F.R.S., Cambridge, U.K.
E. W. Hart, Ithaca, N.Y., U.S.A.
J. P. Hirth, Columbus, Ohio, U.S.A.
E. Krempl, Troy, N.Y., U.S.A.
R. S. Rivlin, Bethlehem, Pa., U.S.A.
J. H. Wiener, Providence, R.I., U.S.A.
M. L. Williams, Pittsburgh, Pa., U.S.A.

This monograph series contains volumes dealing with the mechanical behavior of ceramics, metals, polymers, and their composites. Individual volumes will discuss fundamental as well as applied concepts from both the continuum and microstructural viewpoints. This comprehensive coverage embraces the relation between the phenomenological description of mechanical properties and materials structure.

Fracture Kinetics
of Crack Growth

by
A. S. KRAUSZ

and

K. KRAUSZ
Faculty of Engineering,
University of Ottawa,
Ontario, Canada

KLUWER ACADEMIC PUBLISHERS
DORDRECHT / BOSTON / LONDON

Library of Congress Cataloging in Publication Data

Krausz, A. S.
  Fracture kinetics of crack growth.

  (Mechanical behavior of materials; 1)
  Includes bibliographies and index.
  1. Fracture mechanics. 2. Materials — Fatigue.
I. Krausz, K. II. Title. III. Series.
TA409.K73 1988 620.1'126 87—29725

ISBN-13: 978-94-010-7116-1   e-ISBN-13: 978-94-009-1381-3
DOI: 10.1007/978-94-009-1381-3

Published by Kluwer Academic Publishers,
P.O. Box 17, 3300 AA Dordrecht, The Netherlands.

Kluwer Academic Publishers incorporates
the publishing programmes of
D. Reidel, Martinus Nijhoff, Dr W. Junk and MTP Press.

Sold and distributed in the U.S.A. and Canada
by Kluwer Academic Publishers,
101 Philip Drive, Norwell, MA 02061, U.S.A.

In all other countries, sold and distributed
by Kluwer Academic Publishers Group,
P.O. Box 322, 3300 AH Dordrecht, The Netherlands.

To the memory of
Henry Eyring

Table of contents

Preface

Over the past few years, we have made numerous presentations, delivered several series of lectures, and participated in many discussions on the processes of time-dependent crack growth. We felt that the understanding of these processes had reached a degree of maturity: the basic physical principles were established and their application to engineering practice was now feasible. We concluded that the best way to organize this knowledge was to write it up in a single, coherent system. Martinus Nijhoff kindly encouraged us and generously offered their collaboration. Hence, this book.

The physical process of time-dependent subcritical crack growth is rigorously defined by statistical mechanics. If well presented, the principles can be readily understood by practitioners of fracture research and design engineers. We present the physical processes of crack growth in terms of atomic interactions that assume only a working knowledge of the standard engineering materials course contents. From this, we develop a framework that is valid for any type of material, be it metallic, polymeric, ceramic, glass or mineral — indeed, any solid. We also assume an elementary exposure to fracture mechanics. An appendix is provided that outlines those aspects of fracture mechanics that are needed for an introduction to fracture kinetics analyses; it also provides a common ground for concepts and terminology (see Appendix A).

We proceed through theory to applications that are of interest in research, development and design, as well as in test and operating engineering practice. Many applications are worked out in detail; others are left to the reader to elaborate. To encourage our readers to apply theory to particular cases, specialized mathematics is usually avoided. Where this was not possible, the mathematical apparatus is introduced in sufficient detail, always bearing in mind the standard mathematical tools of the engineering profession.

We admit to partiality in our presentation. We believe that other approaches are also possible, but all of these must be compatible. Nature has many faces, but only one character. In saying this we are restating, in the context of crack growth, the general principles of the exact sciences. To serve these principles we had to leave unreported much valuable, even important, work. It was a painful decision, a difficult selection process, and we ask the forbearance of those who would have chosen otherwise. We also wanted, for now, to limit the size of the book, since a more extensive work is in progress.

The book is intended to be used as a textbook in graduate courses; as well as by researchers and design and test engineers. To facilitate the extension of the use of fracture kinetics, references are listed for each chapter; we want to draw particular attention to the books that are referred to most often. These will serve the reader well.

The advanced practice of fracture control, be it engineering research, design or operation, utilizes the principles of rationally derived damage tolerance analyses. This book is based on this principle, and contributes to its application.

Each chapter starts with simple concepts that are developed into increasingly sophisticated and complex particular cases. For all chapters, theory is presented first, followed by applications. We subscribe to the argument that real understanding is attained only when concepts appear so simple they can be 'touched', and that this can only be achieved by direct involvement, by doing. We thus strongly advise the reader to work through the examples.

We expect that when this knowledge is attained, the reader will experience the pleasure one always feels on perceiving how clever Nature is: she does all with elegance and simplicity. We expect that the reader's professional practice will benefit as well.

We want to express our gratitude to Sherman Brown, George Hahn and Max Williams for their advice and comments. These were most valuable and contributed much. We want to thank Mary Fraser for her competent editing and Denise Champion-Demers for the careful typing: we owe them much for their kind and unfailing patience with us. The figures were prepared by Kamlesh Gupta; we admire his skill and imagination. To John Mshana, and to the good number of student and professional audiences on whom we honed the book — our appreciation.

One of us (A. S. Krausz) thanks the Natural Sciences and Engineering Research Council of Canada for financial assistance.

A. S. Krausz
K. Krausz

List of symbols

a	crack size
a_0	initial crack size
a	interatomic distance
a_0	equilibrium interatomic distance
C, n	experimental, fatigue constants
C^*	C^*-integral
E	Young's modulus; modulus of elasticity in tension or compression
ΔE	energy (of the Arrhenius equation)
$\mathcal{G}$	crack-extension force
$\Delta G^+(W)$	apparent activation energy
ΔG^+	activation energy
h	Planck's constant
J	J-integral
$\mathcal{k}$	elementary rate constant
k	Boltzmann constant
K	stress intensity factor
L	crack size growth per activation $\times$ environmental effects
N	number of cycles
R	load ratio
S	strain energy density function
T	temperature; Kelvin scale
t	time
v	velocity
W	mechanical energy
α	work factor parameter of fracture mechanics
Δ	increment
ε	normal strain
ρ	number of cracks
σ	normal stress

	Subscripts
b	breaking
c	critical
d	degrading-step related
h	healing
i	counter
ISCC	threshold stress intensity in SCC
I, II, III	Regions I, II, III of SCC
j	counter
r	reversal of degrading step
th	threshold

The physical background of time- and temperature-dependent crack growth

Empirical fracture relations are important tools in engineering practice but they provide information only within the region of measured and tested behavior: extrapolation from these regions using empirical equations is dangerous. Reliable predictions on crack growth behavior can be made only if they are based on physical principles: the establishment of rational, quantitative, crack growth relations is essential to advanced engineering design and testing procedures. These relations are formulated in terms of external and internal constraints. They define the crack velocity as the function of

(a) external constraints: the load and displacement boundary conditions, the component and crack geometry, the chemical and thermal environment, and their variation in time;

(b) internal constraints: the microstructure of the material; chemical composition, the spatial arrangement of the constituent atoms (such as the structure of crystalline materials), the type, configuration and distribution of defects, and their variation in time.

The factors that control crack velocity, and their effects, must be expressed in exact quantitative forms derived rigorously from physical principles. Fracture kinetics is dedicated entirely to the satisfaction of this need. It combines the theories and practice of thermal activation and fracture mechanics. Accordingly, time-dependent crack growth behavior is defined explicitly as

rate of crack growth $= f$(load, displacement, geometry, thermal and
chemical environment, microstructure, and their
evolution over time).

Chapter 1 reviews briefly those aspects of the physical processes of slow crack growth that form the background of fracture kinetics. The review begins with the classical Griffith theory, then continues with the description of the physical processes of crack growth on which fracture kinetics is based. This framework will be developed further into an exact mathematical formulation — the rate theory of all time- and temperature-dependent processes. It will be shown that rate theory expresses quantitatively the effects of the microstructure, the load, and temperature distribution, and the effects of their changes during crack growth. This is followed by a summary of how the Griffith theory has been extended and modified to provide a cornerstone for the modern understanding of crack growth. The entire chapter establishes the foundation for the theory and application of fracture kinetics and the development of constitutive laws.

1.1. The Griffith theory

By the 1920s, it was recognized that materials are much weaker than their calculated atomic strengths would indicate. Griffith raised the important questions of why is this so, and how the real fracture strength might be determined [1−14].

The theoretical strength of a perfect crystalline material can be calculated approximately by considering the simple atomic structure shown in Figure 1.1. Figure 1.2 represents the interatomic force versus interatomic distance for each atom pair. Under the effect of a tensile force, the interatomic distance increases from a_0 to $a_0 + \Delta a$; the change can be expressed in terms of the normal strain as $\varepsilon = \Delta a / a_0$, and hence $\Delta a = a_0 \varepsilon$. The stress, σ, is proportional to the interatomic force and can be approximated as

$$\sigma = \sigma_{\text{crit}} \sin \frac{2\pi}{\ell} \Delta a. \tag{1.1}$$

(See Figure 1.2 for the meaning of ℓ.) Because in elastic deformation the change in interatomic distance is small, the stress—strain relation (the force—distance relation of Figure 1.2) is linear, in agreement with the Taylor expansion of the sinus function

$$\sigma = \sigma_{\text{crit}} \frac{2\pi}{\ell} \Delta a = \sigma_{\text{crit}} \frac{2\pi}{\ell} a_0 \varepsilon.$$

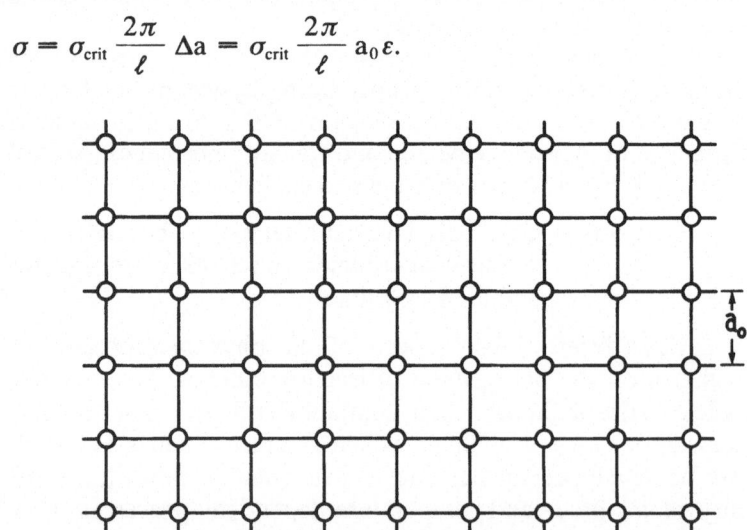

Figure 1.1. The arrangement of atoms in a simple crystalline structure. The distance between two atoms when no force is acting is a_0, the equilibrium distance.

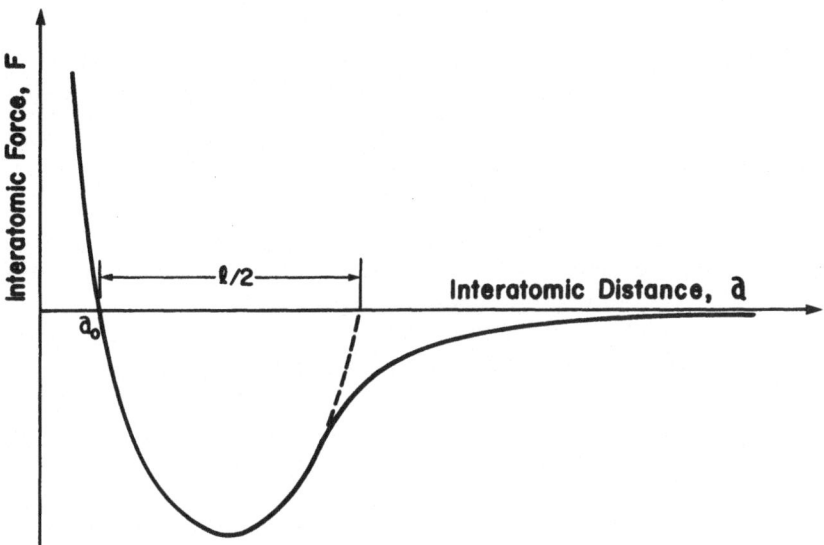

Figure 1.2. The solid curve shows the force versus distance relation between two atoms; the dashed curve represents the sinusoidal approximation, with wavelength ℓ. See also Figure 1.7.

The slope of the linear stress—strain relation provides the elastic modulus E:

$$\frac{d\sigma}{d\varepsilon} = \sigma_{\text{crit}} \frac{2\pi}{\ell} a_0 = E. \tag{1.2}$$

From equation (1.2) the critical stress, at which the atoms separate until fracture occurs, is

$$\sigma_{\text{crit}} = \frac{E\ell}{2\pi a_0}.$$

Assuming further, as a first approximation, that $a_0 = \ell/2$, the critical stress is of the order

$$\sigma_{\text{crit}} = \frac{E}{\pi};$$

more accurate calculation shows that

$$\sigma_{\text{crit}} \cong \frac{E}{10}. \tag{1.3}$$

Because, in reality, materials are so much weaker than equation (1.3) would predict, Griffith's conclusion was inevitable: the model is incomplete, and defects in the material must be present that reduce its strength.

To determine the size of the defects that cause fracture at the actual stresses, Griffith suggested an energy argument. He considered that crack growth is accompanied by the creation of new surfaces, increasing the energy content of the solid. This energy must be supplied: its sources are the work performed by the applied force and the stored elastic energy. The corresponding energy balance is:

change in surface energy = change in work + change in stored elastic energy.

A crack can grow only if enough energy is available. Consider, for illustration, the simple test with a specimen containing a through crack, loaded first in tension and then fixed at both ends so that its length remains constant, as in Figure 1.3(a). At fixed length the load performs no work: the energy for crack growth is supplied entirely by the change in the stored elastic energy.

Inglis showed that in an infinite plate of unit thickness with an elliptic through crack, such as that shown in Figure 1.3(b), the elastic energy change is

$$-\Delta U_e = \frac{\pi \sigma^2 a^2}{E},$$

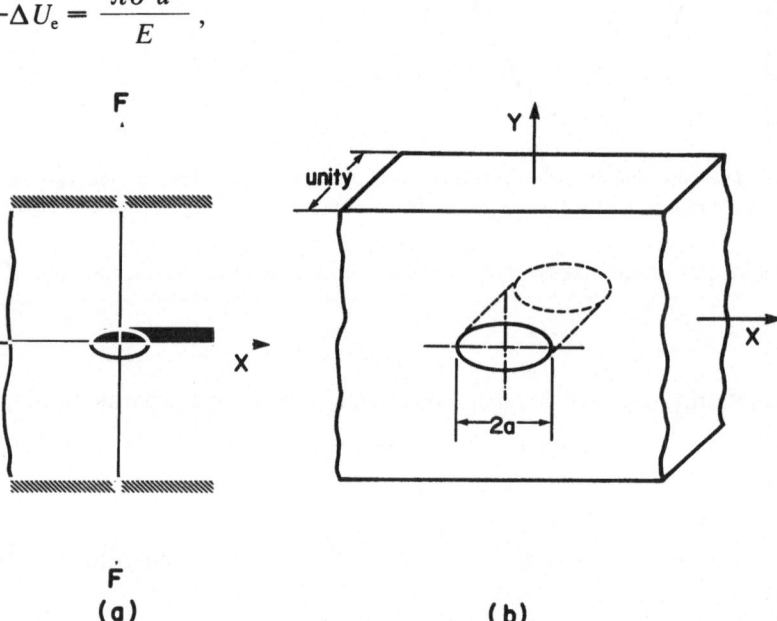

$$(a) \qquad\qquad\qquad (b)$$

Figure 1.3. (a) The stressed state of a plate with fixed ends; (b) the schematic representation of an infinite plate of unit thickness with a through crack.

where a is the half-crack size. The surface energy change is

$$\Delta U_s = 4\gamma a,$$

where γ is the specific surface energy, and because two new surfaces are created at each crack edge, the factor 4 is used for a crack size of $2a$. The energy condition is defined by

$$\Delta U_{total} = \Delta U_e + \Delta U_s$$

$$= -\frac{\pi \sigma^2 a^2}{E} + 4\gamma a, \qquad\qquad (1.4)$$

as shown in Figure 1.4. When the crack is of the critical size a_c, the energy equilibrium is unstable. A crack of size $a_c - da$ shrinks because the system can thus reduce its total energy, $\Delta U_e + \Delta U_s$; crack of $a_c + da$ size grows spontaneously to fracture because this also reduces the energy of the system.

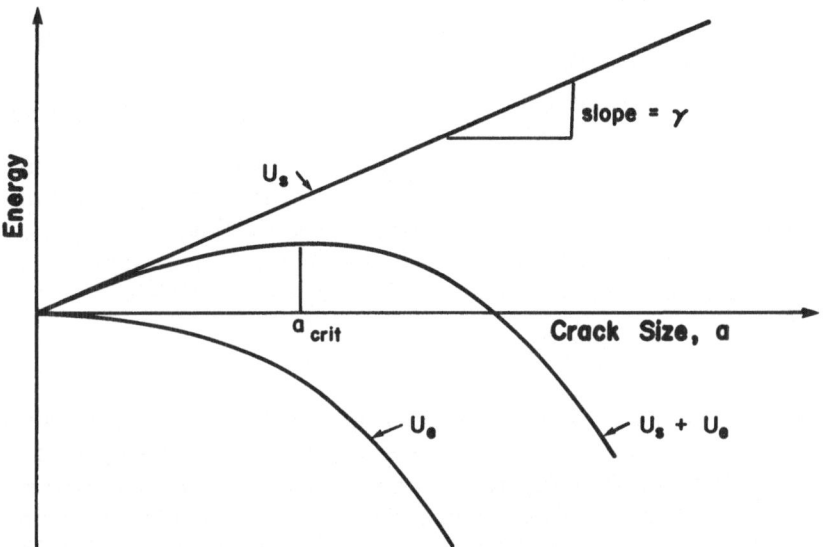

Figure 1.4. The variation of the surface energy U_s, the stored elastic energy U_c, and the total energy $U_s + U_c$.

The critical stress that brings the crack to the unstable equilibrium state is defined by the condition

$$\frac{d\,\Delta U_{total}}{da} = 0;$$

and, from equation (1.4),

$$\frac{d}{da}\left(\frac{\pi\sigma^2 a^2}{E} - 4\gamma a\right) = 0$$

the critical stress is

$$\sigma_{crit} = \left(\frac{2\gamma E}{\pi a_c}\right)^{1/2}.$$

The Griffith theory has had considerable impact on further developments, despite certain limitations imposed by the state of knowledge at that time. Subsequently, it was established that cracks and other defects are always present in materials, causing reduced fracture resistance; this understanding is now a cornerstone of fracture mechanics. The theory also led to the recognition that the stress level at which fracture occurs varies inversely with the critical crack size. Furthermore, it established that crack growth in a linear elastic body is controlled by two material characteristics: the elastic moduli and the surface energy. The theory anticipated also that cracks can heal back under specific conditions, and suggested that crack growth occurs quickly at a critical velocity, leading to catastrophic fracture.

Experiments on elastic solids have partly supported these conclusions. The energy argument of the Griffith theory — that crack growth requires energy, and the energy supply controls growth — is certainly valid for all time. This concept will be elaborated further in later chapters.

The Griffith theory is, however, limited to linear elastic continuum. In materials where plastic deformation also occurs, as in ductile metals, the energy absorption of the plastic flow can be much greater than the surface energy, and the energy required to propagate the plastic zone controls crack growth.

Orowan and Irwin proposed the energy balance

$$4\gamma a + 2\Gamma a = \text{energy supplied},$$

where Γ is the plastic energy needed for the crack to grow by a unit distance. When large-scale plastic flow accompanies the crack growth, the plastic energy is much greater than the specific surface energy, and the latter is negligible.

The Griffith theory has required other modifications and extensions: some of these are noted later in this chapter. Nevertheless, it provided the foundation for the modern theory and practice of crack propagation control — which is, of course, the ultimate goal of all fracture-related activities. Since the 1950s, further advances have established fracture mechanics as the theory that defines the mechanical state of a crack. The theory of fracture kinetics is relatively new — one can locate its origins in the 1960s, or even later. Fracture mechanics and fracture kinetics are both developing rapidly, and their frontiers are still open.

1.2. Review of the physical processes

The competent use of fracture constitutive equations requires considerably more than the use of ready-made formulae: it is a creative, problem-solving endeavor. The practitioner must be able to see past the mathematical machinery describing fracture processes and perceive the underlying physical conditions that control them. The development and application of new engineering techniques, and conceptual advances in the science of fracture control, must be based on an understanding of the submicroscopic processes: all the secrets of time- and temperature-dependent phenomena are embedded at this level. Armed with this knowledge, engineers engaged in research, design, testing and maintenance can determine the mechanisms of crack growth through rational, rather than empirical, means.

The theory of fracture kinetics is derived from this understanding of the atomic-level physical processes that control crack growth. Familiarity with the behavior of materials at this level, at least to the extent and depth elaborated in this review, is thus essential for anyone engaged in the control of crack growth and fracture failure. This section offers a review that assumes the usual, common undergraduate engineering background in physics needed for the application of advanced damage-tolerant design principles in the design and operation of structures and machinery. The presentation is simplified for expediency, at some sacrifice to precision. Generally speaking, this will not detrimentally effect engineering practices or compromise research interests. However, readers requiring a more detailed development for some research purposes are provided with the references. A brief review of the concepts of fracture mechanics that are needed for the understanding of fracture kinetics is given in Appendix A.

This discussion begins with a description of the fundamental processes of crack growth, shared by crystalline and amorphous materials. Using these universal concepts, the practitioner can investigate any type of material without continually reorienting basic premises to address new conditions.

How cracks grow

Consider a crack in a plate that grows under the effect of an applied load. If this load is sufficient, growth occurs virtually instantaneously, at about two-thirds the velocity of sound. When the load is significantly less than critical, slow, subcritical crack growth takes place: Figure 1.5 represents a typical behavior. Two important questions then arise. What is the physical cause of the subcritical crack growth? and How does the velocity depend on the macroscopic and

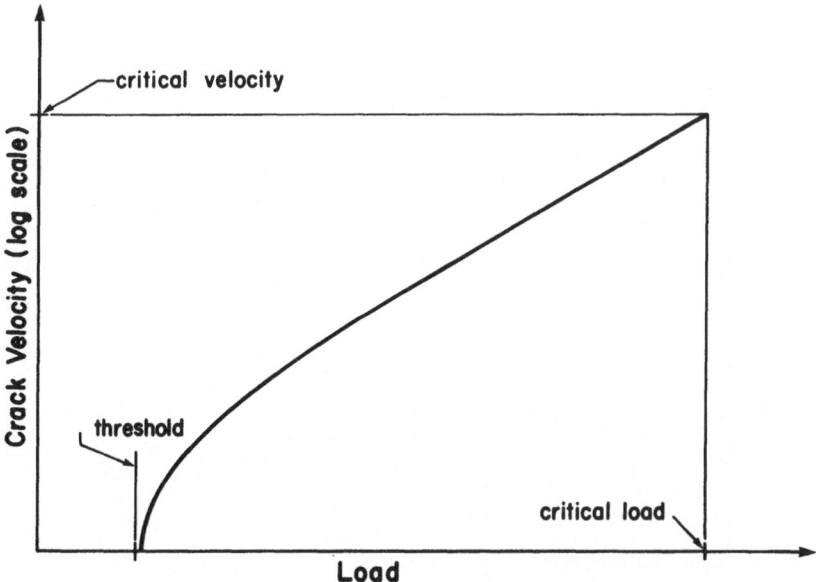

Figure 1.5. Schematic representation of the typical crack velocity dependence on the load at constant temperature and microstructure.

microscopic effects? Both queries can be answered within the context of fracture kinetics.

Figure 1.6(a) gives the usual schematic representation of a plate with a crack growing in the x-direction. To understand the physical mechanism, processes must be considered at the atomic scale, where crack velocity is controlled. Figure 1.6(b) shows schematically the atomic arrangements at the crack-tip

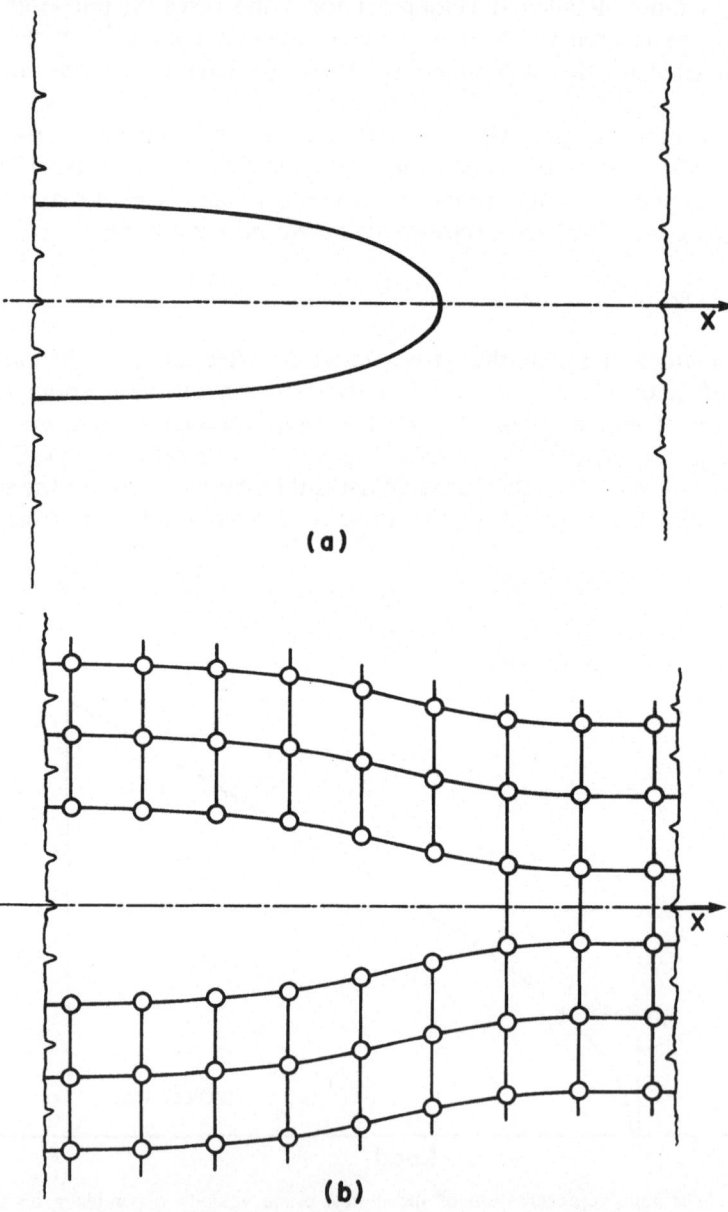

Figure 1.6. (a) Schematic representation of a crack in a solid continuum model; (b) the same crack-tip region considered at the atomic level.

zone. The circles represent the atoms of a simple structure, and the lines indicate that interatomic forces act between the atoms, holding the solid together. Missing lines signify that the bonds, and thus the coherence of the solid, are broken, producing a crack. Crack growth results from a succession of such bond-breaking steps.

This simple observation underlines the fundamental process of crack growth in any solid under any conditions. It is a universally applicable principle, and provides a basis for extensive further developments.

Figure 1.7(a) represents the interatomic force that acts between two atoms. As the force applied to the solid is increased, the interatomic distance grows. When the applied force reaches maximum, the atom pair separates, as shown in Figure 1.7(a)*. Consequently, when an applied force reaches this maximum at the crack tip, the bond breaks and the crack moves ahead by one atomic distance.

Under a sustained load this step is repeated continuously, and the crack grows until fracture occurs. The crack velocity is limited only by the usual physical constraints associated with the growth of mechanical effects: the maximum possible crack velocity in a specific solid is about one-quarter to two-thirds the velocity of sound. It should also be noted from Figure 1.7 that

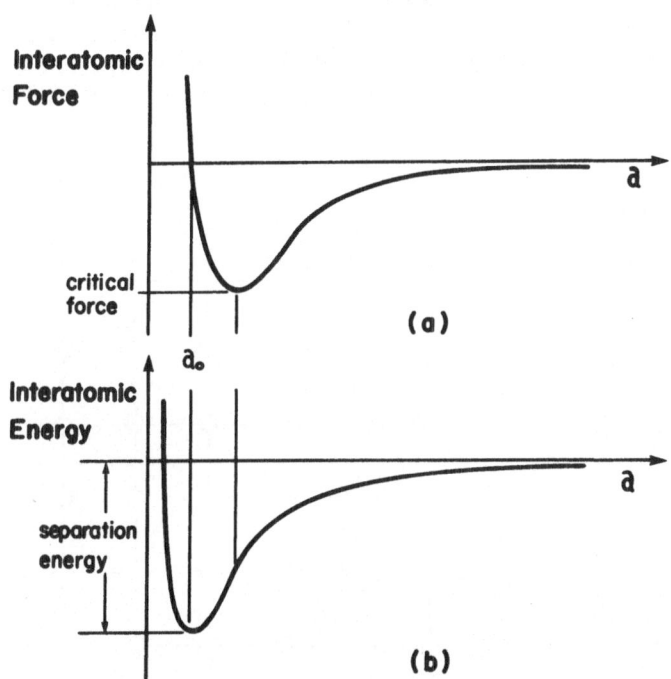

Figure 1.7. (a) The interatomic force versus distance relation; (b) the interatomic energy versus distance relation. Note that, by definition, force = −d energy/da; (b) shows this accordingly.

---

* Note that the applied force is equal in magnitude but opposite in sign to the interatomic force.

the progression satisfies the energy conditions. Figure 1.7(b) is the integrated curve of Figure 1.7(a):

$$\text{interatomic energy change} = -\int_{a_0}^{a} \text{force } da,$$

where a_0 is the equilibrium atomic distance.

The energy required to move the crack by one atomic distance — to break one bond — is supplied by the available work, which consists of the applied work and the released elastic strain energy. This model then satisfies the conditions of critical crack propagation velocity as established by the Griffith theory; extends the theory by accounting for the real atomic structure of the material; and gives a clear definition of the surface energy, γ, of the Griffith theory

$$\gamma = \frac{\text{separation energy}}{a_0^2}.$$

However, like the Griffith theory, this model is still incomplete because it does not explain subcritical crack growth. Figure 1.8 illustrates the atomic configuration in the crack-tip zone. Again, lines between the atoms indicate the interatomic forces, while the circles represent the atoms at the positions they occupy under the effect of the applied load. But, of course, the atoms are never stationary: they vibrate continuously about the positions indicated by the circles. The vibrations are at a very high frequency, in the order of $10^{12} \, s^{-1}$ and with

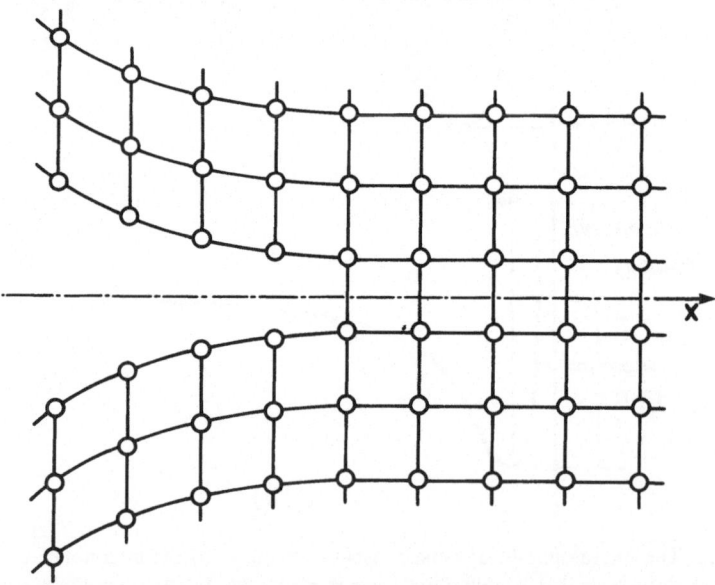

Figure 1.8. The configuration of the atoms at the crack-tip zone.

randomly changing amplitudes and directions. This behavior is integral to the process and must be incorporated in the model in order to complete the representation of time- and temperature-dependent crack propagation. It is this physical model that leads to the fracture constitutive laws [8—13].

To illustrate how the vibration of atoms affects crack propagation, Figure 1.9 represents schematically one layer of the atomic structure of a solid. The atoms are shown as small, vibrating masses interconnected with springs. The random vibrational energy is perceived on the macroscopic scale as thermal energy, the heat content of the material. Because a solid contains atomic masses in the order of 10^{24} cm^{-3}, this behavior cannot be described by the usual differential equations of the vibrating many-body systems: statistical methods must be applied. To this end, extensive numerical analyses have been carried out in which atomic interaction effects were simulated to model crack growth processes [14].

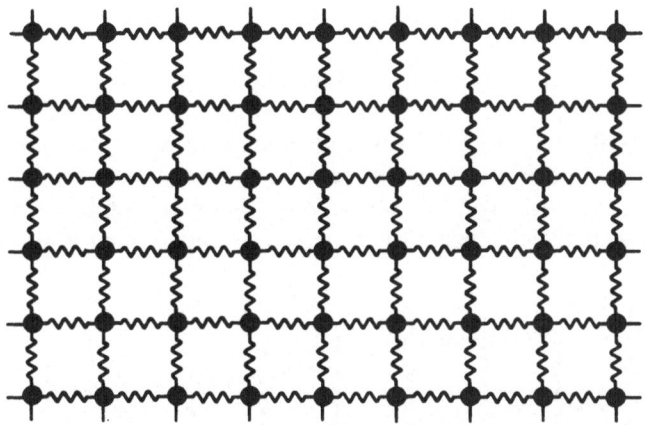

Figure 1.9. The model of solids that allows for the vibrational behavior of the constituent atoms, represented by the dots as 'heavy masses'.

It has been shown that, in this type of system, the vibrations of individual masses constantly change in amplitude and direction. These changes propagate through the solid and are perceived as a wave motion. This is illustrated in Figure 1.10. The open circles and interconnecting light lines represent the atoms at their equilibrium positions, when they are at a distance a_0 apart. The vibrations move them from these positions. The configuration of atoms at a particular instant is shown by the solid circles and the heavy lines. At the next instant, as the wave crest moves to the next atom, the configuration is changed. Notice that in some places the atoms are pulled much further apart than the equilibrium distance, resulting in greater interatomic energy. Random changes in interatomic distances as the waves move are thus accompanied by random changes in interatomic energy. This random, wavelike fluctuation of energy controls — indeed, produces — subcritical crack growth.

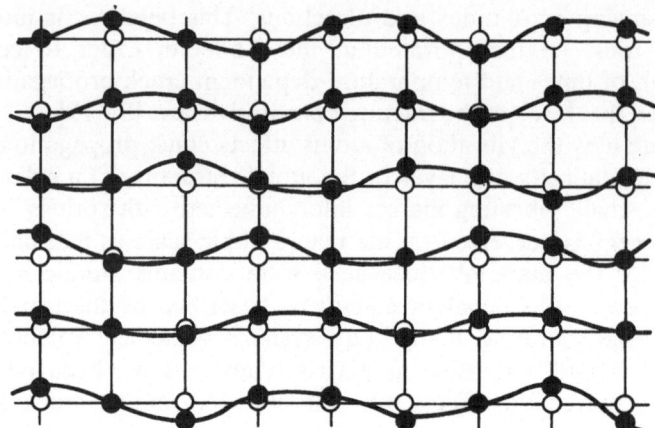

Figure 1.10. An instantaneous state in the atomic configuration as the vibrational motion sweeps through the solid. The open circles represent the atoms at their equilibrium position, while the solid circles indicate their 'actual' positions.

Consider again the crack-tip zone, focusing this time on the model in Figure 1.11. Here, the atoms are represented as being interconnected by springs that allow them to vibrate. Atoms at some distance from the crack tip vibrate about their equilibrium position; near the crack tip the atoms are displaced from equilibrium — the 'springs' are stretched. Energy is already stored in these springs, that is, between these atoms. When a load is then applied the atomic

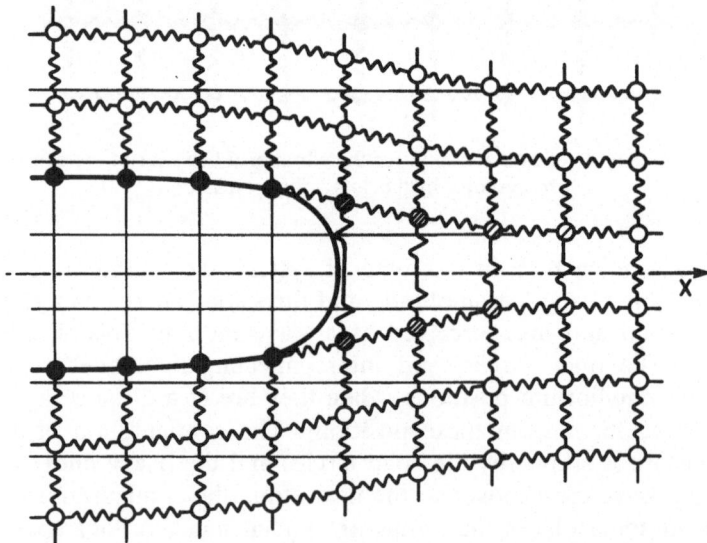

Figure 1.11. Schematic representation of the crack-tip region. The model allows for the vibrational motion of the atoms. Springs indicate that atoms are interacting; missing springs indicate loss of interaction — broken bonds. The region without 'springs' is equivalent to the free surface — that is, the crack surface — and is shown by the heavy line.

distances are increased. As noted before, if the load is sufficiently large to increase the interatomic force to the critical level (Figure 1.7), the bond breaks immediately and the crack moves by one atomic distance. When the applied load is insufficient for bond breaking, the displaced atoms continue to interact: the 'springs' in front of the crack tip are pulled, but do not break.

Without vibration, the stretched configuration would be static. However, due to the vibrations, the interatomic distance between the pair of atoms just ahead of the crack tip fluctuates, as shown by the cross-hatched circles in Figure 1.11. As changes in amplitude sweep through the material with a wavelike motion, at some instant a crest will arrive that is large enough to separate the atoms to the extent that the 'spring', or atomic interaction, breaks. The atomic configuration after this event is shown in Figure 1.12. Comparison with Figure 1.11 shows that the atomic configuration in front of the newly positioned crack tip is the same as the previous one: the process of crack growth is repeated step by step.

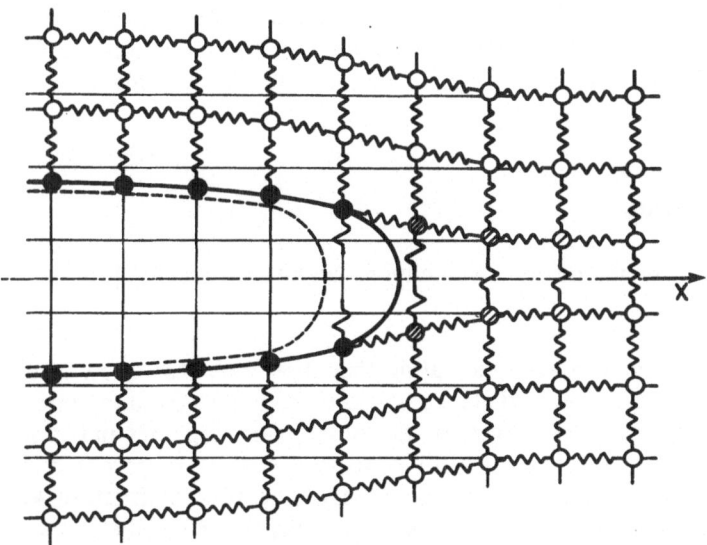

Figure 1.12. The atomic configuration after the crack has moved by one atomic distance. The broken line shows the previous location of the crack.

It is emphasized that this model of a crack propagation step is very much simplified. In reality, the atomic configuration is three-dimensional and the crack tip propagates along a crack front. However, the description is sufficient for much of what follows; where refinements are needed to account for the three-dimensional configuration, it will be extended appropriately.

The waves associated with the changes of interatomic distances are not harmonic functions. Figure 1.13 illustrates their random character, which results from irregularities associated with the presence of material defects and from other causes. Thus, changes in atomic distances are randomly distributed and, consequently, wave crests large enough for bond breaking will reach the crack

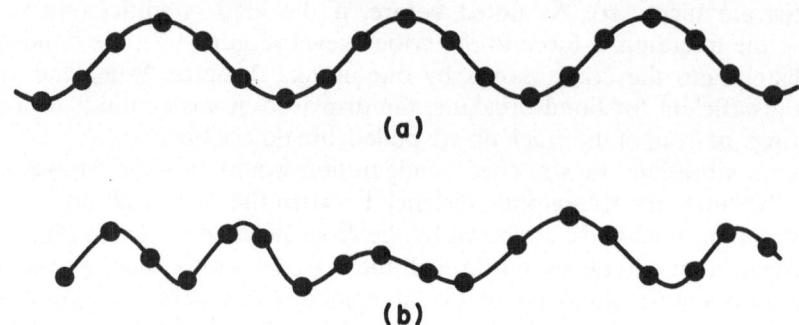

Figure 1.13. (a) The idealized, harmonic-function-type wave of a row of atoms as represented by Figure 1.10; (b) the more likely, random, waveform of the atomic vibration propagation.

tip at random intervals. This random effect is of the utmost importance, as it controls probabilistic time- and temperature-dependent crack growth.

To establish the quantitative expressions of the constitutive law of fracture, this model must now be expressed mathematically. To this end, the theory of rate processes is introduced.

1.3. Rate theory

In its simplest form, the crack velocity, v, can be expressed as [6]*

$$v = na_0 \mathcal{k}$$

where n is a small integer indicating that the crack may grow at each step by one or more interatomic distance a_0; and $\mathcal{k}$, the elementary rate constant as defined by rate theory, represents the number of steps per unit time. It will be shown later that all crack growth processes can be represented by the combination of the elementary rate constants. The task of fracture kinetics is to establish the appropriate combination that fully describes the crack propagation process under investigation.

The widely applied theory of rate processes describes the behavior of the atomic configuration that results from the random fluctuation of thermal energy. All such processes are essentially the same and are described identically by the rate theory. The most familiar of these are the transport processes (mass transport by diffusion, heat transport by conduction, momentum transfer in viscous substances); chemical reactions; and the time- and temperature-dependent plastic deformation of solids. Together with time- and temperature-dependent crack growth, these constitute a single category. This is a critical observation because, as Chapter 2, Part 2, and Chapter 3, Part 2, demonstrate, some of these processes are often associated with, and contribute to, subcritical crack growth [6, 15—21, 28—30].

* References to Chapter 1 are given on page 137.

The discussion of the physical processes of crack movement showed that atoms vibrate with random amplitudes and in random directions, and that when a wave of sufficient amplitude reaches the atom pair at the crack tip their bond breaks. From Figure 1.14 it follows that a change in the distance between two atoms corresponds to a change in the atomic interaction energy. A high-amplitude wave crest carries with it a high energy content. This randomly distributed and varying energy is, in fact, the heat content of the solid, the thermal energy, and the fluctuation of the interatomic distance is the local fluctuation of the thermal energy. The frequency of the distance change, k, must thus be described in terms of energy.

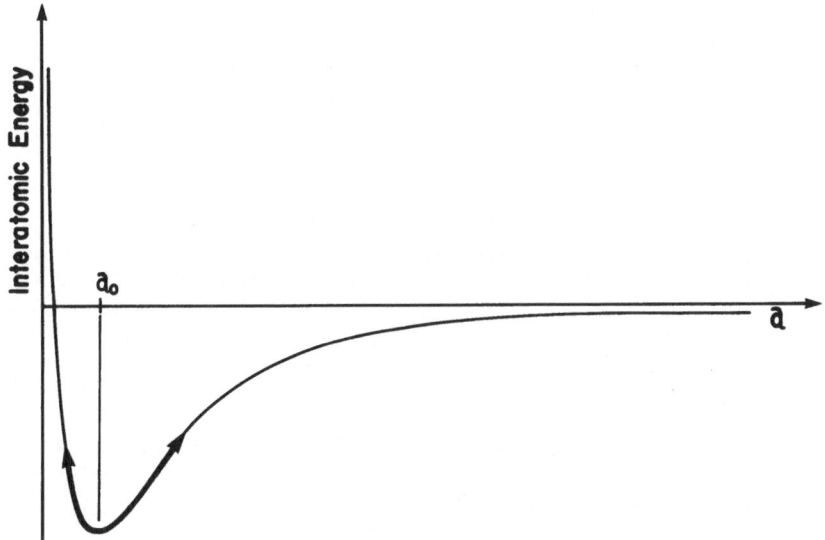

Figure 1.14. The arrow represents the changes in the interatomic distance as the atoms vibrate about the equilibrium distance a_0.

Because the number of atoms in a solid is very large, statistical means must be used to express the mechanical and thermal effects. The fundamental theory of statistical thermodynamics (or, as it is alternatively called, statistical mechanics) is employed to derive the elementary rate constant. The following brief description, which assumes only a background common to all engineers, is sufficient for most applications to fracture control. For those with theoretical and research interests, the derivation of the rate constant is presented in Appendix B, in the context of the transition state theory.

Detailed analyses demonstrate that when a crack grows by one step the atomic interaction energy changes as shown in Figure 1.15. At a given crack size, at a particular instant, the energy associated with the configuration of atoms is as low as possible: the system is in an energy valley. The size of the crack grows by the steps of na_0. For this to occur, the energy must be increased first by the amount ΔG_b^+. The configuration change then continues, with

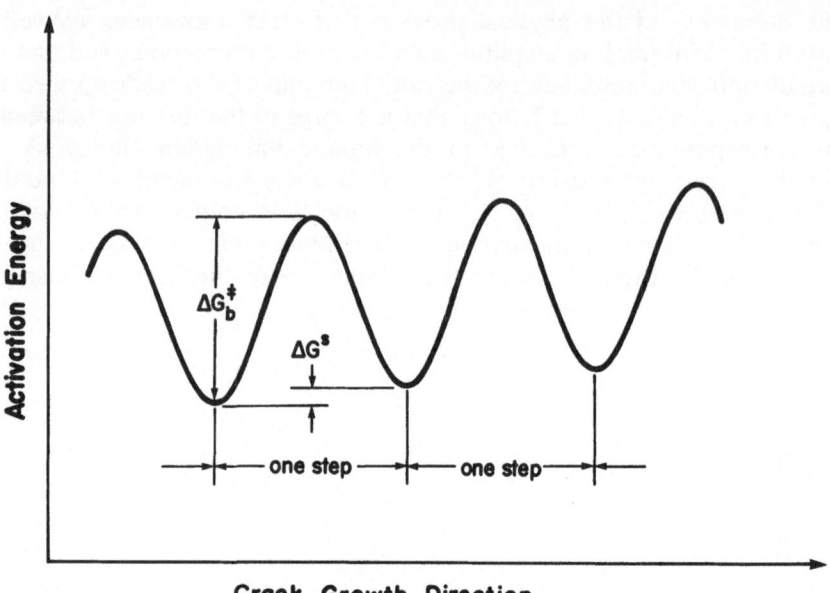

Figure 1.15. The schematic representation of the energy change as the crack grows by steps of na_0.

decreasing energy, down into the next valley: the crack has grown by one step. This process repeats itself as the crack propagates. In brittle fracture, the total energy of the atomic system increases after each step by the appropriate amount of surface energy, ΔG^s. The rate of overcoming the energy barrier, the frequency of steps, is the rate constant. This is expressed by the rate theory as [6]

$$k_b = \frac{kT}{h} \exp\left(-\frac{G_b^+ - W_b}{kT}\right), \tag{1.5}$$

where the subscript 'b' signifies that the quantity is associated with the atomic bond-breaking process, with crack growth. The elementary rate constant, k, is the number of steps per unit time; k and h are Boltzmann's and Planck's constants respectively, both of which are universal constants ($k = 1.38 \times 10^{-23}$ JK^{-1}; $h = 6.62 \times 10^{-34}$ Js); T is the absolute temperature; and W is the work that contributes to the rearrangement of the atomic configuration while the crack grows from the energy valley to the top of the energy barrier. This work is provided by the external load and by the change in the stored elastic energy [8–13].

The elementary rate constant, k, is a fundamentally important quantity. All time- and temperature-dependent crack growth processes are described as the functions of k, which is expressed in terms of well-defined physical quantities and represents all effects that control crack growth. The activation energy, ΔG_b^+, is associated with the rearrangement of the atoms: it thus represents quantitatively the effects of the microstructure. The work, W, depends on the forces that drive the crack growth. In accordance with the concepts of fracture mechanics it is expressed, for linear elastic solids, as the function of the stress intensity factor, K, or the crack-extension force, $\mathcal{G}$; for non-linear elastic or elastic—plastic bodies, as the function of J-integral, or the strain energy function S; and, for crack growth in time- and temperature-dependent plastic solids with C^* or $\dot{J}$. It was noted early in this chapter that these characteristic quantities are described in terms of the load and geometry conditions. The rate theory considers that the work depends on these quantities,

$$W = f(K; \text{ or } \mathcal{G}; \text{ or } J; \text{ or } S; \text{ or } C^*; \text{ or } \dot{J})$$

as well as on the material characteristics. Accordingly, W represents the load and geometrical boundary conditions and those microstructural characteristics that are associated with the contribution of the work to the crack growth step. The corresponding design factors are rigorously and quantitatively represented, and it is important to note that the rate constant also describes explicitly the effect of the temperature.

The quantity $\Delta G_b^+ - W_b$ in equation (1.5) has special significance. Its physical meaning is explained as follows.

To produce a crack growth step, the energy of atomic rearrangement, ΔG_b^+ (true activation energy), must be present. This is only partly available from the work; until the energy deficiency of $\Delta G_b^+ - W_b = \Delta G_b^+(W)$ is supplied, the system cannot surmount the energy barrier represented in Figure 1.16, and the

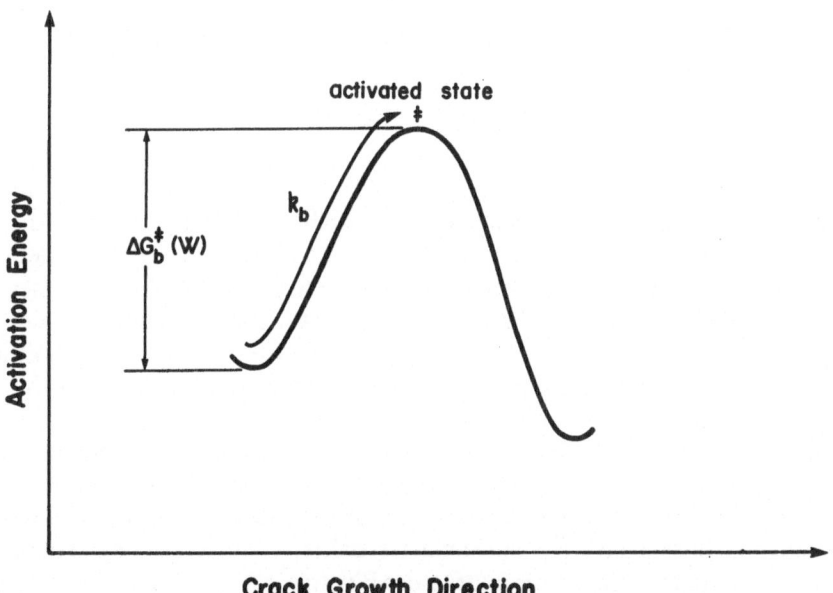

Figure 1.16. The energy barrier of a crack growth step. The state that corresponds to the energy peak during the step is called the activated state.

crack cannot grow. It was noted above that the thermal energy is the energy associated with the random vibrational motion of the atoms, and that as this motion propagates, high-amplitude, high-energy wave crests reach the crack-tip zone at random intervals. These wave crests supply the thermal energy (apparent activation energy) $\Delta G_b^+(W)$; together with the work, they provide the energy needed to rearrange the atoms: $\Delta G_b^+ = W_b + \Delta G_b^+(W)$.

Because these wave crests arrive at random intervals, the waiting time in front of the energy barriers is also random. The frequency of steps, called the frequency of activation, is an average quantity: the elementary rate constant, k_b, is the average number of activations over the energy barrier per unit time. This randomness is the underlying phenomenon that leads to the probabilistic behavior of crack propagation.

Statistical mechanics states, as a universal principle, that atomic systems change not only in one direction, but regroup in the reverse direction as well. In crack growth, this means that the crack tip occasionally shrinks back by one step. This occurs when vibrations of appropriate direction and sufficient magnitude reach the crack-tip zone, producing interatomic distances that bring the atoms close enough to reverse the process shown in Figure 1.11 and 1.12 [22].

This healing process has been amply demonstrated. As early as the 1930s, researchers showed that even macroscopic crack healing can be produced under certain conditions. Numerous investigations have confirmed this observation: a variety of this effect is now recognized as one of the fundamental processes of fatigue. These considerations are discussed further in later chapters. Here, only the single-step healing process at the atomic level is considered [8—13, 23—29].

It follows from rate theory that the frequency of the backward activation, i.e. of the healing step, is

$$k_h = \frac{kT}{h} \exp\left(-\frac{\Delta G_h^+ + W_h}{kT}\right),$$

where the symbols have the same meaning as in equation (1.5), and the subscript 'h' indicates that the quantity is associated with the healing step. The theory recognizes also that the activation energy and work in a healing step may differ from those of a bond-breaking step. Note that with bond healing the energy balance is $\Delta G_h^+(W) = \Delta G_h^+ + W_h$; the thermal energy, $\Delta G_h^+(W)$, must not only supply the activation energy but also counterbalance the work that now *resists* the occurrence of the step.*

Figure 1.17 represents the breaking and healing energy concepts. It is to be noted that usually (with important exceptions) $k_b > k_h$. Consequently, on the macroscopic scale, only crack growth is observed. An interesting and important condition for design and lifetime analysis exists when $k_b = k_h$. These considerations are part of the theory of fracture kinetics.

---

* For further details see Figure 2.2 and Appendix B. For most applications, a basic background in fracture mechanics, and the concepts discussed here, are sufficient.

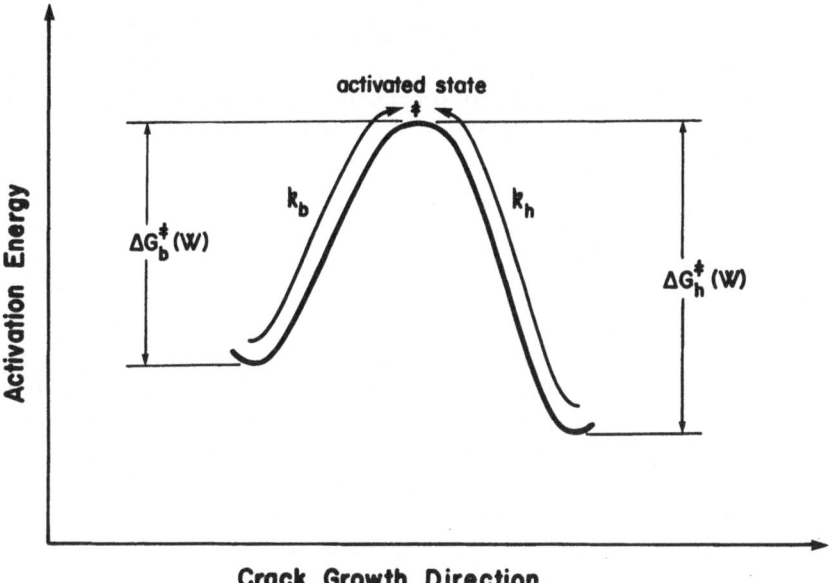

Crack Growth Direction

Figure 1.17. The energy barrier when both bond-breaking and bond-healing occur. Note, in comparison to Figure 1.16, that here the apparent activation energies $\Delta G_b^{\ddagger}(W) = \Delta G_b^{+} - W_b$ and $\Delta G_h^{\ddagger}(W) = \Delta G_h^{+} + W_h$ are represented.

1.4. Comments and summary

The Griffith theory established that in linear elastic solids the propagation and healing of cracks is controlled by the energy balance conditions: i.e. changes in

surface energy < applied work + elastic energy,

and

surface energy > applied work + elastic energy,

respectively.

It also determined that defects must be inherent in the material, otherwise the critical stress would be equal to the theoretical strength. The existence and role of inherent defects are now among the basic elements of fracture mechanics, and advanced design, maintenance and test engineering concepts.

The two conditions noted above are necessary for both crack growth and healing. However, they are not sufficient to define its essential character. Thus, although the Griffith theory provided a cornerstone for later developments, the following modifications and extensions have proved necessary. Concerns listed in (1) to (4) are investigated in the context of fracture mechanics, while those noted in (2) to (6) are addressed through the atomic and statistical mechanics concepts of fracture kinetics.

(1) The linear elastic model is a good approximation of some materials and fracture conditions. Often, however, the non-linear elastic or the plastic models are valid.

(2) While the Griffith theory appropriately focused attention on physically defined material characteristics, the surface (or plastic-zone) energy is very difficult to measure, and to develop practical testing methods other relevant characteristics must be defined.

(3) While the surface energy reflects some aspects of the material properties, the state of the microstructure ahead of the crack tip is a significant factor. For example, the formation of a string of voids ahead of the stress peak in ductile materials affects the crack growth process profoundly.

(4) Although the overall energy conversion condition remains valid as a thermodynamic principle, fracture mechanics and fracture kinetics recognize that a global energy balance condition is insufficient for crack growth; energy with sufficiently large force component must be available at the crack tip, where the activities occur.

(5) Fracture kinetics recognizes that cracks often grow slowly, in a time-dependent manner, rather than according to the 'all-or-nothing' condition of the Griffith model.

(6) In recognizing the time-dependent nature of crack growth, the fracture kinetics theory takes rigorous account of the essential role of the temperature on subcritical crack velocity.

Although this list seemingly indicates strong deficiencies in the Griffith model, the major contribution of the theory is beyond question. Fracture mechanics is rooted in the Griffith theory, while the energy-based theory of fracture kinetics is similarly in firm agreement with the principles of Griffith.

Fracture mechanics, then, uses practical engineering tests to provide overall quantitative measures of the effects of the material properties. It gives an exact and reproducible description of the stress and strain state; it seeks to represent the effects of the material on crack velocity. However, because fracture mechanics is a continuum mechanics theory, it cannot be used to enquire into the complex physical processes at the microscopic level where crack growth is controlled.

Fracture kinetics incorporates fracture mechanics while considering crack growth where it actually occurs, and at the discrete atomic level. It recognizes that cracks grow when the interatomic distance at the crack tip increases to the point where the atoms cease to interact. This results from the joint action of the work (the applied load and the change of stored elastic energy) and the randomly fluctuating thermal energy of the solid. Therefore, crack growth

(i) is composed of discrete steps of one atomic distance, or an integer multiple of the atomic distance;

(ii) occurs at discrete, randomly distributed time intervals;

(iii) is controlled by the following energy balance of bond breaking:

$$\text{energy needed} = \text{energy supplied},$$
$$\Delta G_b^+ = W_b + \Delta G_b^+(W)$$

and of bond healing:

$$\text{energy needed} = \text{energy supplied},$$
$$\Delta G_h^+ + W_h = \Delta G_h^+(W)$$

The average frequency of steps is defined by the transition state theory of rate processes for bond breaking as

$$\mathscr{k}_b = \frac{kT}{h} \exp\left(-\frac{\Delta G_b^+ - W_b}{kT}\right),$$

and for bond healing as

$$\mathscr{k}_h = \frac{kT}{h} \exp\left(-\frac{\Delta G_h^+ + W_h}{kT}\right).$$

The process is, of course, time-dependent: the frequency of steps defines the rate of propagation. It is also temperature-dependent. This dependence is expressed explicity through $\Delta G^+(W)$, the thermal energy peak, and kT, the average thermal energy of the solid. The physical model and the corresponding rate theory description are typical for thermally activated processes.

The constitutive equation expresses the crack velocity in terms of the load,

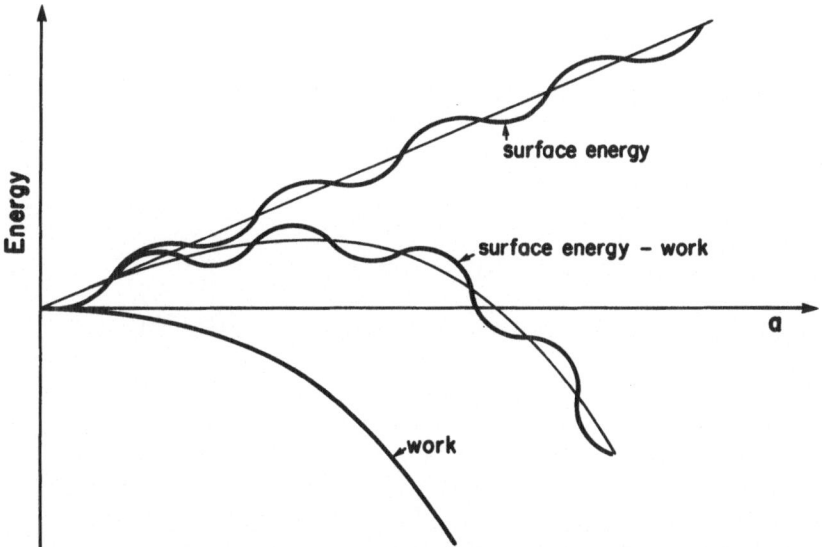

Figure 1.18. Representation of energy conditions according to the classical Griffith theory (light lines) and a simplified form of the revised theory (heavy lines). The revised theory considers that the surface energy is a function of the discrete steps involved in crack growth and the corresponding total energy. Note that the work is continuous in both the solid continuum model and the model that considers atomic structure.

geometry, temperature, displacement, microstructure and changes in these factors, as

$$v = g(\mathscr{k})$$

because $W = f(K$, or $\mathscr{G}$, or J, or S, or C^*, or $\dot{J}$ and the microstructure); the g and f functions represent the external constraints and the microstructure explicitly and quantitatively. Fracture kinetics thus provides the theory and the experimental methods to determine the fracture constitutive laws of time- and temperature-dependent, thermally activated, crack velocity at subcritical levels.

The classical and revised fracture kinetics forms of the Griffith theory are compared in Figure 1.18. As the surface energy—work curve indicates, crack growth actually occurs over a system of consecutive energy barriers, with the activation energy increasing and then decreasing. This behavior is discussed in the context of the following chapters.

Deterministic fracture kinetics theory and constitutive laws

PART 1
DETERMINISTIC FRACTURE KINETICS THEORY

Subcritical crack growth is usually more complex than a sequence of single bond-breaking steps. All mechanisms consist of various processes; and these processes may include diverse steps, each represented by the appropriate expression of the elementary rate constant [1, 3, 4]*:

$$k_i = \frac{kT}{h} \exp \left(- \frac{\Delta G_i^* \pm W_i}{kT} \right). \tag{2.1}$$

Obviously, to control crack growth one must first determine the specific combination of steps that compose the mechanism of fracture, and describe these steps in terms of the appropriate rate constants. Fracture kinetics provides the method to achieve this, through the derivation of the constitutive equation that expresses the complex mechanism of crack growth.

The method is more readily visualized when it is considered, according to the terminology of fracture kinetics, as a means of expressing a specific combination of rate-controlling energy barriers. It classifies the wide variety of possible mechanisms into two categories: processes that can be represented adequately by the effects of a single rate-controlling barrier; and processes that must be represented by several energy barriers, in systems that may be parallel (independent), consecutive (dependent) or a combination of both. This classification provides a satisfactory description of almost any crack growth mechanism.

2.1. The energy-barrier model of fracture kinetics

The simplest fracture kinetics model represents a single energy barrier. Figure

* References to Chapter 2, Part 1, are given on page 138.

2.1 shows that the crack moves forward over an energy barrier by one or at most a few atomic distances at the bond-breaking rate $\mathscr{k}_b$, and that it may occasionally heal back over the same barrier at the rate of $\mathscr{k}_h$. The corresponding rate constants are expressed as [1—7]:

$$\mathscr{k}_b = \frac{kT}{h} \exp\left(-\frac{\Delta G_b^+ - W_b}{kT}\right)$$

and

$$\mathscr{k}_h = \frac{kT}{h} \exp\left(-\frac{\Delta G_h^+ + W_h}{kT}\right).$$

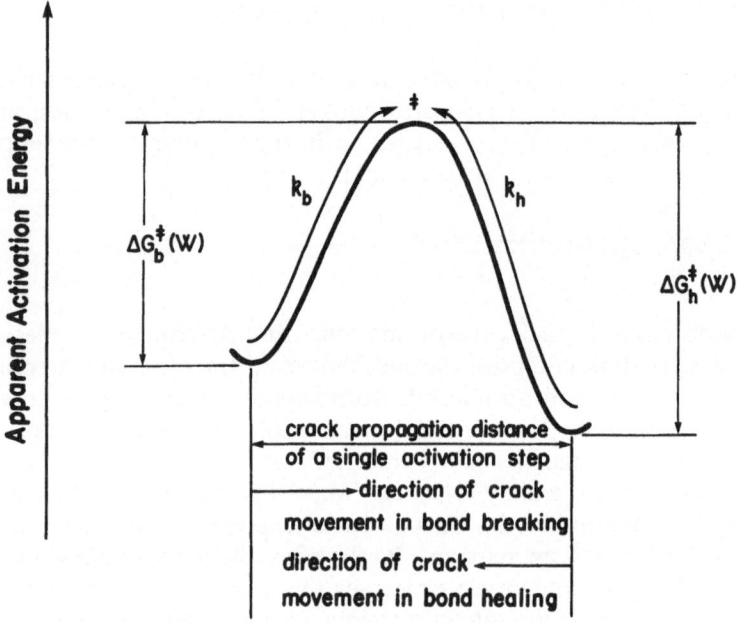

Figure 2.1. Single-energy-barrier kinetics. The subscripts 'b' and 'h' signify that the quantity is associated with the breaking and the healing step, respectively.

These expressions differ only in the sign of the work term. As Figure 2.2 shows, this is because in bond-breaking the applied force helps the crack move forward, while in healing it hinders it from moving backward. This energy bookkeeping, as pointed out in Chapter 1 and illustrated in Figure 2.3, is [6—8]:

(a) for a bond-breaking step

$$\text{energy needed} = \text{energy supplied}$$
$$\Delta G_b^+ = W_b + \Delta G_b^+(W);$$

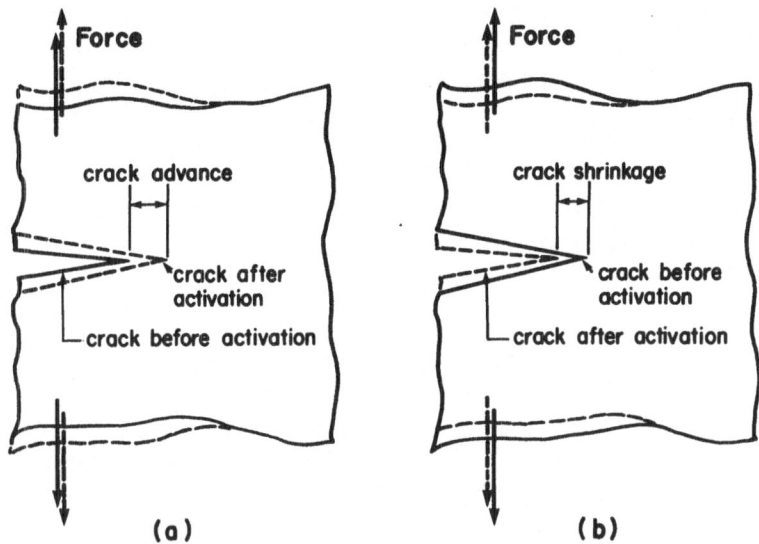

Figure 2.2. A schematic representation of the force—crack movement geometry conditions. (a) The dashed lines represent a bond-breaking step. (b) The dashed lines represent a bond-healing step. The figure demonstrates that during bond breaking the applied force contributes work to the crack movement step, while during healing it absorbs energy.

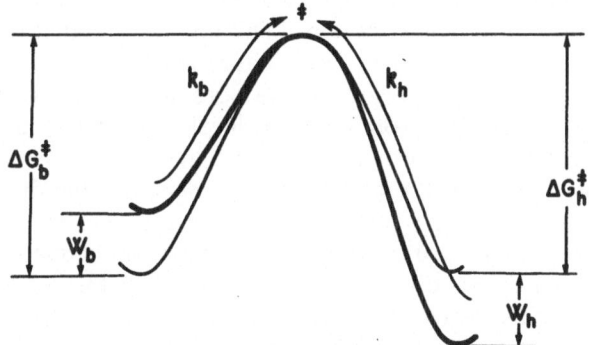

Figure 2.3. The energy barrier and the energy balance condition for bond-breaking and bond-healing steps. The light line indicates the energy barrier without the applied work; the heavy line represents the barrier on application of the work.

(b) for a bond-healing step

$$energy\ needed = energy\ supplied$$
$$\Delta G_h^{\ddagger} + W_h = \Delta G_h^{\ddagger}(W),$$

because energy is needed to overcome the force opposing the activation of the healing step, rather than promoting it, as in breaking. The subscripts 'b' and 'h' signify that breaking and healing may not reflect the same bond energy change and work. Under special conditions, the energy barrier is symmetrical so that $\Delta G_b^+ = \Delta G_h^+$ and $W_b = W_h$ can be observed.

The fracture constitutive equation is the resultant of the two steps:

$$v = L_b \ell_b - L_h \ell_h; \tag{2.2}$$

or again, when the energy barrier is symmetrical ($\Delta G_b^+ = \Delta G_h^+$ and $W_b = W_h$), as, for example, in diffusion-controlled processes, the velocity relation equation (2.2), using equation (2.1), reduces to

$$v = L(\ell_b - \ell_h)$$

$$= L \frac{kT}{h} \left[\exp\left(-\frac{\Delta G^+ - W}{kT}\right) - \exp\left(-\frac{\Delta G^+ + W}{kT}\right) \right],$$

where L is the distance over which the crack moves after each activation.

It is a fundamental principle of statistical mechanics, and consequently of fracture kinetics, that at the atomic level the processes always fluctuate backwards and forwards. The macroscopically observed progression of crack growth simply results from a relatively greater breaking rate; the net effect is, of course, the difference between the two. When both are equal, the net effect is zero and no macroscopic change is observed [9—11].

This phenomenon is one of the mechanisms of the threshold effect. It is often desirable, for design purposes, to establish the maximum value of the facture mechanics parameter at which a crack will still not grow — that is, the threshold value of the parameters K_{th}, $\mathcal{G}_{th}$, J_{th}, etc. Failure to observe crack growth at the macroscopic level over a long period of time does not imply that the crack tip is at rest at the atomic level. Far from it; at this level segments of the crack-tip line flucturate, but because backwards and forwards movements occur at the same frequency, no net change in the crack size is observed. This fluctuation of the crack-tip line has far-reaching practical consequences, as will be shown later.

When first encountered, the concept of backward activations, or crack-healing steps, may present difficulties. To understand this effect, it is useful to consider certain published observations. As early as the 1920s, several investigators demonstrated that under favorable conditions it is possible to produce visible, measurable crack-healing. Recently, more extensive investigations of crack propagation and healing effects demonstrated that the conditions of macroscopic healing are strict and clearly defined: geometrical rematching of the atomic-level structure is necessary, and absence of crack surface contamination and crack-tip plasticity must be assured. Because these stringent requirements can be satisfied only by careful experimental techniques, macroscopic crack-healing is not often observed, and the concept is alien [12—22].

On the other hand, healing on the atomic scale is quite common. When it is recalled that backward activation occurs over one or at most a few atomic distances, it becomes obvious why the requirements for healing that are so difficult to satisfy on the macroscopic scale are often well assured at the atomic level. The constraints of the atomic network promotes matching when the two crack surfaces approach, and the atomic vibration itself may bring the atoms into interaction even if plastic flow has occurred. This is well demonstrated in cases of fatigue, where rebonding is one of the essential components of the mechanism. Even if a surface becomes contaminated, healing can occur when the frequency of the rebonding steps equals or surpasses the contamination reaction rate. Sometimes contamination itself can promote healing on the atomic scale [9—13].

Because their effects accumulate, these healing steps can have a strong influence at the macroscopic level. They therefore have important engineering consequences. Healing is particularly significant in the threshold zone of stress corrosion cracking, mechanical fatigue, corrosion fatigue, and creep rupture processes, as will be discussed later [23].

As stated before, almost all fracture processes are complex combinations of many diverse steps. It is one of the cardinal principles of fracture kinetics that the simplest possible model must always be used to represent a process, and non-essential steps or combinations of steps eliminated. Therefore, a single barrier with bond breaking only should first be adopted as the model. If, however, this proves insufficient, a single barrier with both breaking and healing rate constants provides the next best candidate for the fracture constitutive equation. In the first case, the equation is:

$$v = L k_b, \tag{2.3}$$

and in the second,

$$v = L_b k_b - L_h k_h.$$

If both models prove unsatisfactory, a multi-barrier mechanism must be considered.

Multi-barrier kinetics

The multi-barrier kinetics analysis should start with a two-energy-barrier model. The corresponding equation is usually sufficient to describe a particular behavior; the two-barrier model should be extended to a greater number of barriers only if it clearly proves inadequate.

Two energy barriers can form a parallel or a consecutive system. Sometimes, with experience, inspection of the measured behavior reveals which of the two controls the process; if not, either model may be considered initially valid and subjected to kinetics analysis. The constitutive equation derived from this analysis must then be tested against the observed behavior. Should it be contradicted — and usually the disagreement is very evident — the alternative model must be considered. Examples of this procedure are given in Part 2.

It is emphasized here that the decision on the number of barriers involved is not an artifact for mathematical curve-fitting purposes as it is in the empirical equations, but a necessary consequence of the real, physical conditions underlying the processes. Thus, the number of rate constants required to formulate the constitutive equation provides valuable information on the mechanisms of crack growth, and can be used for fracture control.

Systems with two parallel barriers. Often, the mechanism consists of two independent processes; the corresponding kinetics analysis describes the behavior that results when two parallel energy barriers control crack growth.

Parallel processes can be visualized from the analogy shown in Figure 2.4. The resultant of the two flows, each occurring at a different rate, is their sum. When the flow is considerably heavier from one of the containers, the very small contribution of the other can be disregarded. Independent, parallel processes of crack propagation can be considered in much the same light [24].

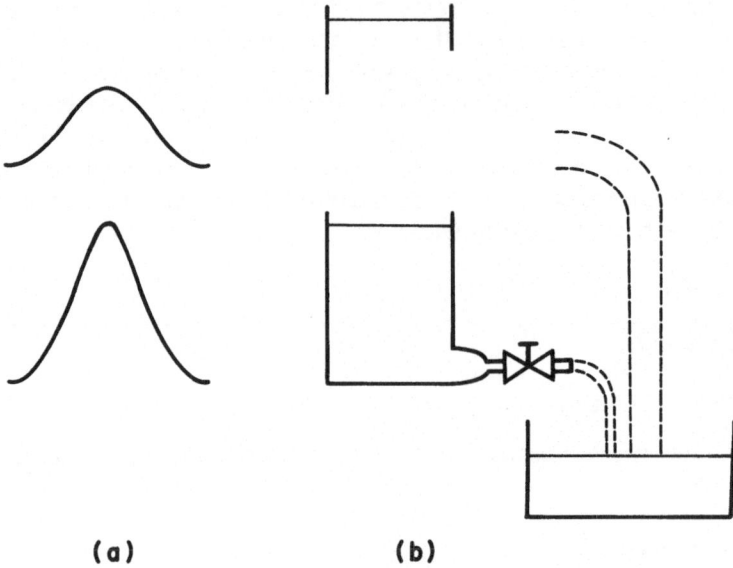

(a)   **(b)**

Figure 2.4. (a) Schematic representation of a system of two parallel energy barriers; (b) the hydraulic analogy of the system. Because the flow rate of one line is independent of the flow rate of the other, the total contribution of the two flows is their sum.

The contribution of each parallel mechanism to observed crack growth is represented in Figure 2.5. The net flow over each barrier is $k_b - k_h$ and the fracture constitutive equation can be written as the sum of the contribution by each. Thus:

$$v = L_1(k_{b1} - k_{h1}) + L_2(k_{b2} - k_{h2}). \tag{2.4}$$

Equation (2.4) is a fully rational expression of the fracture constitutive equation. It defines velocity in terms of loading conditions; the geometry of both the component and the crack; the environment, including the temperature; and the microstructure.

Under certain conditions, one of the two terms on the right-hand side of equation (2.4) is so much smaller than the other that it is negligible, and the

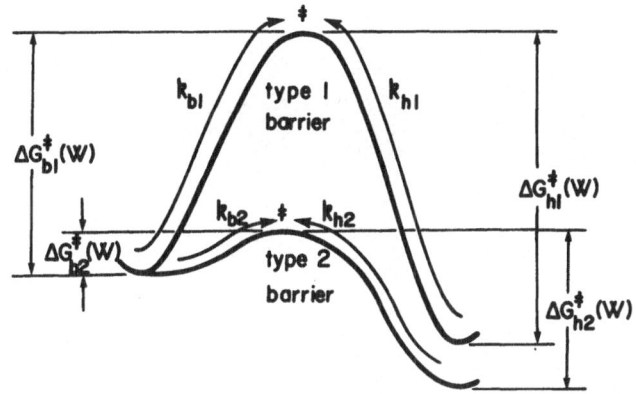

Figure 2.5. The energy—balance condition of two barriers in parallel.

effective result is a single-barrier kinetics. This is why a single-barrier description of the constitutive equation is often adequate; and it is only in this sense that single-barrier kinetics has meaning.

Systems with two consecutive barriers. Mechanisms are often controlled by a dependent, or consecutive, system of two energy barriers, with crack propagation steps occurring in sequence. The analogy for consecutive systems is shown in Figure 2.6, while Figure 2.7 provides an example of crack growth as a consecutive process in a layered composite material [24, 25].

Figure 2.7 shows clearly that the crack must invade a hard layer before it can penetrate the adjacent soft layer, and vice versa: each step can proceed only after the previous one is completed. The kinetics equation, which is somewhat more complicated than for the parallel process, is derived as follows.

Figure 2.8 represents a system of consecutive energy barriers, with two types of barriers recurring in alternating sequence. This discussion will first assume that bond healing is negligible.

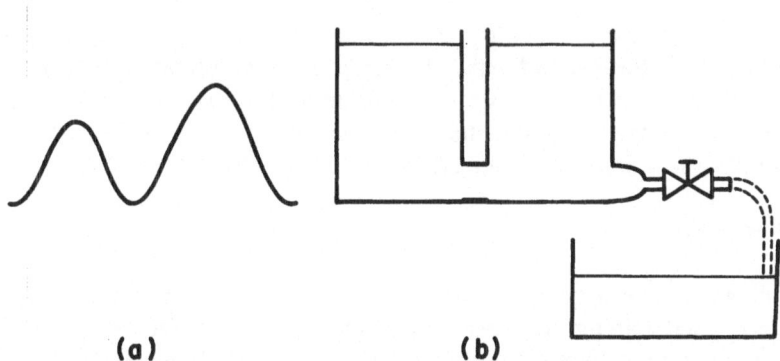

(a)   **(b)**

Figure 2.6 (a) The schematic representation of two energy barriers in sequence. (b) The hydraulic analogy.

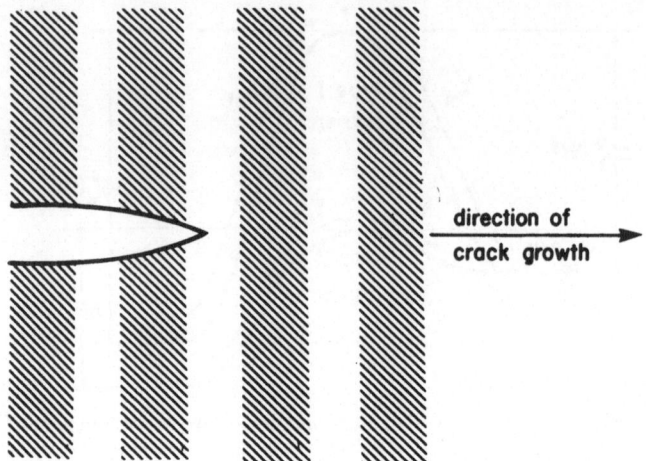

Figure 2.7. Consecutive crack growth in a layered composite material. The shaded and unshaded areas represent hard and soft layers, respectively.

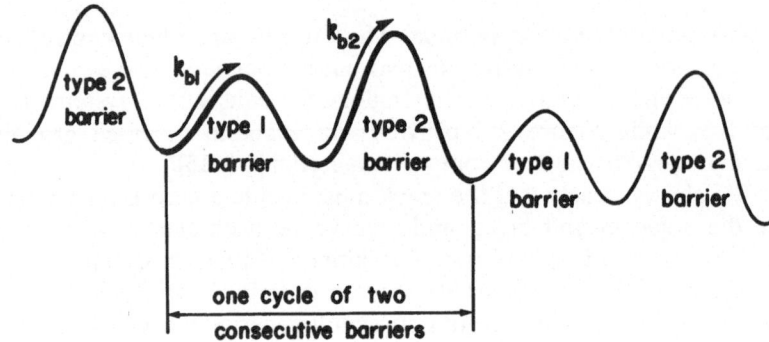

Figure 2.8. A schematic illustration of the two-barrier consecutive system with negligible bond healing.

Over a two-barrier consecutive system, the rate of crack advance is equal to the rate at which the second type of barrier is surmounted. When the healing rate is negligible, the overall rate of overcoming the two-barrier system — the overall activation rate constant, the overall frequency — is the reciprocal of the total waiting time, t, between crack advances [37] (see also Appendix C):

$$k = \frac{1}{t}.$$

The total waiting time is the sum of the individual waiting times t_1 and t_2, in front of each barrier. Because each waiting time is the reciprocal of the corresponding rate constant, or activation frequency, it follows that

$$t = t_1 + t_2 = \frac{1}{k_1} + \frac{1}{k_2} = \frac{k_1 + k_2}{k_1 k_2}$$

and hence,

$$\mathscr{k} = \frac{1}{t} = \frac{\mathscr{k}_1 \mathscr{k}_2}{\mathscr{k}_1 + \mathscr{k}_2} = \frac{1}{\dfrac{1}{\mathscr{k}_1} + \dfrac{1}{\mathscr{k}_2}}. \tag{2.5}$$

Substituting equation (2.1), equation (2.5) is described explicitly as

$$\mathscr{k} = \frac{kT/h}{\exp\left(\dfrac{\Delta G_1^* - W_1}{kT}\right) + \exp\left(\dfrac{\Delta G_2^* - W_2}{kT}\right)}.$$

Here, as for the two-barrier parallel system, the constitutive law of crack growth is expressed fully in terms of microstructure, crack and component geometry, loading conditions, temperature, and environment.

Systems with more than two barriers. For these mechanisms the constitutive equations can be expressed by direct extension of the two-barrier systems [2, 5, 24, 29, 34, 35].

For *m* parallel barriers the constitutive equation is:

$$v = \sum_{j=1}^{m} L_j \mathscr{k}_j = \sum_{j=1}^{m} L_j \frac{kT}{h} \exp\left(-\frac{\Delta G_j^* - W_j}{kT}\right)$$

and for *n* consecutive barriers (Appendix C):

$$v = \frac{1}{\displaystyle\sum_{i=1}^{n} L_i^{-1} \frac{1}{\mathscr{k}_i}} = \frac{1}{\displaystyle\sum_{i=1}^{n} L_i^{-1} \frac{h}{kT} \exp\left(\dfrac{\Delta G_i^* - W_i}{kT}\right)}. \tag{2.6}$$

When *n* consecutive barriers are combined in *m* parallel branches, the constitutive equation is:

$$v = \sum_{j=1}^{m} L_j \left[\frac{1}{\displaystyle\sum_{i=1}^{n} L_i^{-1} \frac{h}{kT} \exp\left(\dfrac{\Delta G_i^* - W_i}{kT}\right)}\right]_j.$$

The kinetics analysis and the interpretation of the constitutive equation follow the concepts discussed for two-barrier systems.

2.2. The principles of the kinetics analysis

The constitutive equation for a specific condition is derived through a kinetics analysis. This must be based on experiments that measure the effects of the external constraints and the material on the velocity, and allow the number, configuration and characteristics of the energy barriers to be established. A quantitative representation is thus obtained in terms of k, the elementary rate constants: the resulting constitutive equation describes rigorously the crack growth behavior for the specific constraints and material under investigation.

The full analysis of the mechanism can be time-consuming and difficult. However, because the constitutive equation is expressed in terms of physically based design concepts — load, geometry, temperature, degrading environment, and material — it can be used with confidence outside the range of the tested behavior, provided that validity limits (such as structural changes due to annealing) are well understood. Indeed, the fracture kinetics approach itself provides a basis for validity-limit judgements.

It is emphasized again that the kinetics analysis yields theoretically rigorous, physically based fracture constitutive equations. Terms, such as the stress intensity factor, are established analytically; on the other hand, the material properties must always be measured. This approach is widely recognized and used in other contexts. Consider, for example, the constitutive equation of the deformation of the linear elastic body. The role of the yield limit and the elastic moduli are rigorously derived. However, their values for each specific material must always be measured according to the methods that conform to the elastic constitutive law. Fracture constitutive equations are formulated on the same principle.

The general form of the constitutive law, then, is expressed in terms of the elementary rate constants as $v = v(k_i)$, where the velocity depends on the consecutive and parallel combinations of several rate constants. When, under specific conditions, the velocity is controlled by a single rate constant, the evaluation is straightforward. For many practical cases, and for certain cases of theoretical interest, the calculation of two rate constants provides satisfactory results. As will be shown, this usually presents no methodological problem.

However, the kinetics analysis required to solve multi-rate-constant mechanisms is considerably more complex: the experimental and analytical difficulties increase almost exponentially with the number of rate constants. The direct analysis in such cases is formidable indeed, and an alternative approach, involving the systematic isolation of single rate constants, must be used. Step-by-step guidance for this procedure is presented below.

The first step of the kinetics analysis

Clearly, the simplest form of the constitutive law is the crack-velocity relation controlled by a single rate constant:

$$v = Lk. \tag{2.3}$$

In keeping with the above-noted principle of the fracture kinetics analysis, the validity of this model should be assumed as a first step; the kinetics is much easier to unravel if the analysis can begin with the simple form of equation (2.3). To determine whether this is possible, tests are carried out for a temperature- and stress-factor regime combination that makes a single rate constant dominant. When equation (2.3) is valid, the velocity—stress intensity factor relation is represented by a straight line in the semi-log coordinate system because*

$$\ln v = \ln\left(L\,\frac{kT}{h}\right) - \frac{\Delta G^{*} - \alpha K}{kT}.$$ (2.7a)

A similar relation is valid, of course, for the other stress factors; in the following description of the fracture kinetics analysis, the stress intensity factor will be used. Equation (2.7a) is represented in Figure 2.9. The evaluation of $\ln(L(kT/h)) - (\Delta G^{*}/kT)$ and α is straightforward: the intercept gives the first, while the slope gives the second. Hence, the full expression of equation (2.3) is obtained.

In general, the mechanism is controlled by a single rate constant over a limited range of the stress factor only. At zero or small values of K, the line representing the $\log v$ versus K relation must, quite obviously, curve asymptotically toward the ordinate (if it does not already deviate from the straight line for

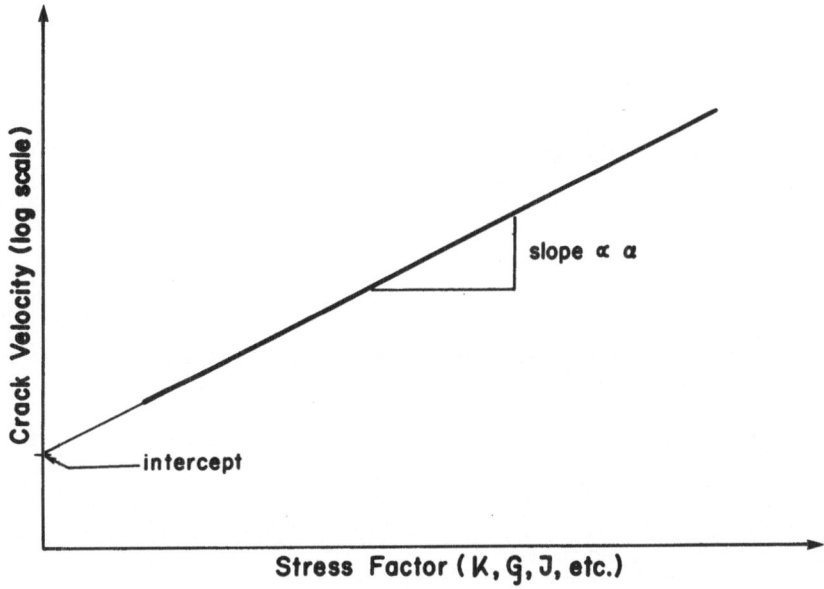

Figure 2.9. The observed behavior of the mechanism controlled by a single energy barrier at constant structure and temperature.

* It is to be noted here that to turn equation (2.7a) into a dimensionless form, it should be expressed as

$$\ln \frac{v}{v_0} = \ln\left(\frac{1}{v_0}\,L\,\frac{kT}{h}\right) - \frac{\Delta G^{*} - \alpha K}{kT}$$ (2.7b)

(where v_0 is a unit velocity factor). However, for expediency, the v_0 factor will not be shown explicitly.

other reasons). The deviation is present because in this region more than one rate constant control the process. This leads to the extension of the analysis to include k_2, the second rate constant.

The second step: the second rate constant

At the next level, the constitutive law contains two rate constants, k_1 and k_2, and the velocity is

$$v = v(k_1, k_2). \tag{2.8}$$

Equation (2.8) has only one unknown: the velocity was measured, and k_1 determined fully, in the first step. The task, then, is to determine the kinetics combination of the two rate constants and to evaluate k_2. This is done through a simple combination of physical and mathematical processing.

From the physical point of view, the second rate constant reveals its effects in the deviation of the curve from the straight line of the log v versus K coordinate system. Fracture kinetics recognizes only three possible associations: (i) k_2 may be a healing rate constant; or (ii) a breaking rate constant in parallel combination with k_1; or (iii) a breaking rate constant in a sequential, consecutive, combination. The associated constitutive equations are

(i) $\quad v = L_b k_b - L_h k_h$

(ii) $\quad v = L_1 k_1 + L_2 k_2$

(iii) $\quad v = \dfrac{1}{\dfrac{1}{L_1 k_1} + \dfrac{1}{L_2 k_2}}.$

To identify the kinetics, collect the known terms (v and k_b or k_1) on the left-hand side, and leave the unknown terms on the right-hand side:

(i) $\quad L_b k_b - v = L_h k_h$

(ii) $\quad v - L_1 k_1 = L_2 k_2;$

(iii) $\quad \dfrac{1}{\dfrac{1}{v} - \dfrac{1}{L_1 k_1}} = L_2 k_2.$

Again, as in Figure 2.9, the left-hand side is plotted on a log scale against the stress intensity factor for each case. Clearly, only one of the three possible kinetics can give a straight-line relation; the other two are represented by curves, and only one of these can have the right sign for the slope. This mathematical consequence identifies the mechanism. One of the trial plots will give the straight line: the intercept is related to the pre-exponential factor and the slope of the line is proportional to the work factor parameter α of k_h or k_2. Thus, the kinetics combination of both rate constants is identified and its full quantitative description obtained.

The third step: the third rate constant

Over an even wider range of stress factor, the effects of a third rate constant will be revealed by a deviation from the straight line in the second plot. The kinetics combination of the third rate constant can be identified by repeating the isolation technique applied in the second step. Write the possible kinetics combinations, and rearrange the corresponding constitutive laws so that all known factors (v, and the two previously determined rate constants) are on the left-hand side; this leaves the third, unknown rate constant on the right-hand side. Plot the logarithm of the left-hand side as the ordinate against the stress factor; only one of the possible kinetics will be represented by a straight line. Then evaluate the intercept and slope to obtain the kinetics identity of the third rate constant and its quantitative description, as was done before.

All further rate constants are identified in the same systematic, step-by-step manner.

The above method determines the effect of the stress factor. It does not, however, resolve the activation energy ΔG^+, which remains embedded in the intercept value $(L(kT/h)) \exp(-\Delta G^+/kT)$. To isolate the activation energy, temperature effects must be considered.

The effects of temperature

Time-dependent crack growth is critically dependent on temperature. This is, of course, a consequence of the physical processes that control subcritical fracture. It is the thermal energy, $\Delta G^+(W)$ that determines the rate of bond breaking in consequence of the energy need, ΔG^+, and the available work, W. Temperature has, therefore, a prominent and explicit place in the mathematical expression of the rate constant; indeed, it appears in both the pre-exponential frequency factor and the exponent. However, it is evident, if typical values are introduced, that the pre-exponential temperature factor is considerably weaker than the effect of temperature in the exponent. Thus, except at values near absolute zero, it is satisfactory to disregard the temperature dependence of the frequency factor and consider the temperature effect only through the exponent.

Because the temperature dependence of the rate constant is exponential, it has a very strong effect on the velocity. Indeed, as a rule of thumb for many practical cases, the rate constant changes by a factor of 2 for every ten-degree temperature change. It is clear, then, why temperature has a key role in the analysis of the mechanism of crack growth.* A valid constitutive equation must accurately reflect both temperature and work dependence, and correspond logically with the physical character of the process as defined in Chapter 1. When these conditions are met, the established constitutive law becomes an important tool for lifetime design and maintenance schedules, and for research and development projects.

The effects of temperature on the rate of thermally activated processes was

---

* Just a reminder: temperature is always expressed in the Kelvin scale.

first investigated by Arrhenius in the late 1880s, and our terminology reflects his findings even today. The Arrhenius equation [1, 3, 36]

$$\text{rate} = A \exp\left(-\frac{\Delta E}{kT}\right) \tag{2.9}$$

represented a major conceptual advance in its time and is still often used.

Equation (2.9) bears a superficial resemblance to the simplified form of the rate constant expression. It is, however, arrived at empirically: neither ΔE, which denotes an energy, nor A, the pre-exponential factor with the dimension of time^{-1}, has real physical meaning. Thus, the equation is incapable of the further extensions allowed by the rate theory, and, because it fails to express clearly and quantitatively the physical meaning of the thermally activated process of subcritical crack growth, its validity is limited. Furthermore, the Arrhenius equation is a single-exponential expression and as such it fails to represent the often complex kinetics; it is an empirical equation. Thus, in the following discussion only the rigorously derived rate constant will be used.

The kinetics analysis of temperature effects starts with the so-called Arrhenius plot, which presents the rate of crack velocity at a constant stress factor and microstructure as the function of the temperature. This has the advantage that in the semi-log coordinate system the constitutive law $v = L\dot{\ell}$ is represented by a straight line, as shown in Figure 2.10. The slope of the line is proportional to the apparent activation energy, $\Delta G^+(W)$, and the intercept of the line with the ordinate provides the numerical value of $L(kT/h)$. Note that because the

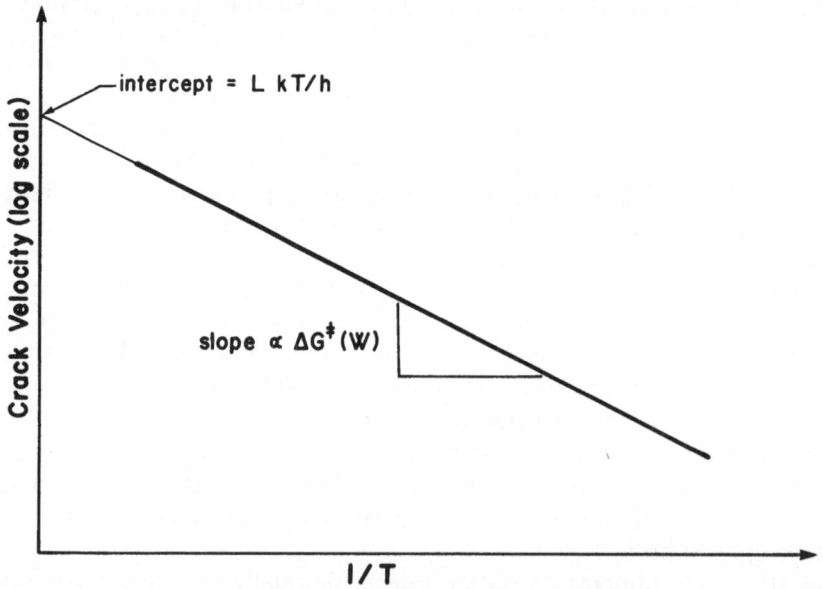

Figure 2.10. The schematic representation of the Arrhenius plot when the crack velocity is controlled by a single rate constant, $v = L\dot{\ell}$.

temperature effect is controlled by the temperature term in the exponent, the temperature in the pre-exponential term has a negligible effect on the straightness of the line.

In the Arrhenius plot direct experimental data are plotted, as measured at constant stress factor and microstructure. The kinetics analysis yields two major insights: the kinetics combination of the energy barriers; and the activation energy ΔG^+ for each of the rate constants, or energy barriers, of the mechanism.

Figure 2.10 represents the simplest kinetics, a process controlled by a single energy barrier with breaking rate constant only. The Arrhenius plot is a straight line; its slope is proportional to the apparent activation energy, and therefore to $\Delta G^+ - W$. Because ΔG^+ is the total atomic-bond energy change associated with the crack-movement step, it is a characteristic value of the material microstructure and expresses the associated, fundamentally important factors quantitatively and rigorously.

It is important to be aware that the slope of the Arrhenius plot, as measured directly from the experimental data, gives only the apparent, or experimental, activation energy, $\Delta G^+(W)$. It is not in itself a representative, characteristic value of the microstructure and the mechanism: it depends on the work, W, and can take on any value between zero and ΔG^+. Thus, to obtain physically meaningful, useful information, ΔG^+ must be evaluated: and, while $\Delta G^+(W)$ is measured with relative ease, the determination of ΔG^+ is quite difficult.

To determine ΔG^+, W must be defined from the velocity versus stress factor plots. Alternatively, it can be derived through a series of measurements employing the Arrhenius-type diagrams (Figure 2.11). The slopes are plotted as the function of K and extrapolated to zero K, where $\Delta G^+(W) = \Delta G^+$.

Analysis of the temperature effects for two-barrier kinetics

Again, the simplest possible kinetics model should be assumed, and its validity tested. When the mechanism clearly consists of more than one rate constant, a kinetics analysis must be carried out. The following discussion will present an alternative approach, assuming a model of two energy barriers with breaking activation only, that yields the kinetics combination through an inspection method employing the Arrhenius diagrams. In many cases, this time-saving technique is sufficient for systems composed of two rate constants.

For a two-barrier system with breaking activation only, the Arrhenius plot is composed of two intersecting straight lines, each representing a rate constant, as shown in Figure 2.12. In the shaded zones (a) and (b), the rate constants are considerably different, and one of them has only a negligible effect; in the unshaded zone, the two rate constants are of similar value and both affect the crack velocity. Their interpretation for various mechanisms is as follows.

For *two parallel barriers*, the crack velocity is described as

$$v = L_1 k_1 + L_2 k_2$$

and it is then the larger term that dominates the process. In zone (a) this is k_2,

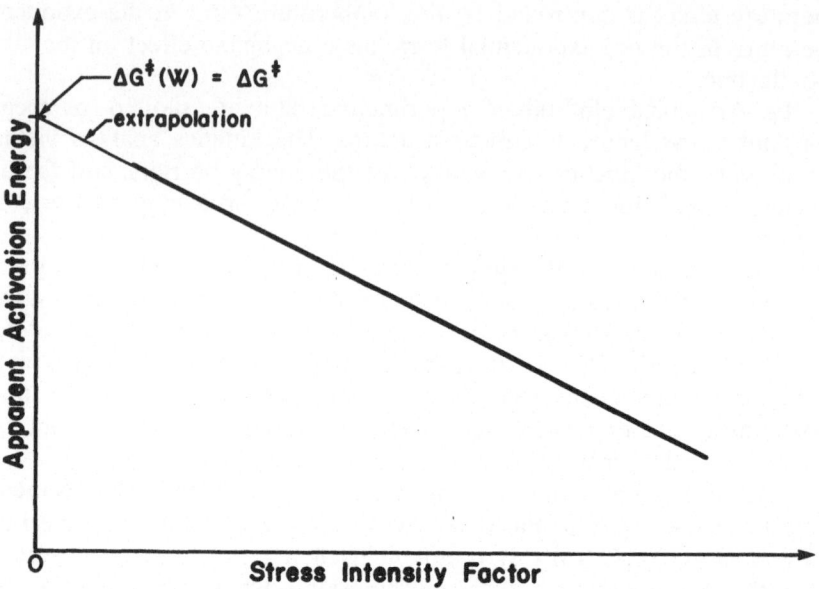

Figure 2.11. The schematic representation of the extrapolation method of determining activation energy. The heavy line was evaluated from the Arrhenius plot, Figure 2.10.

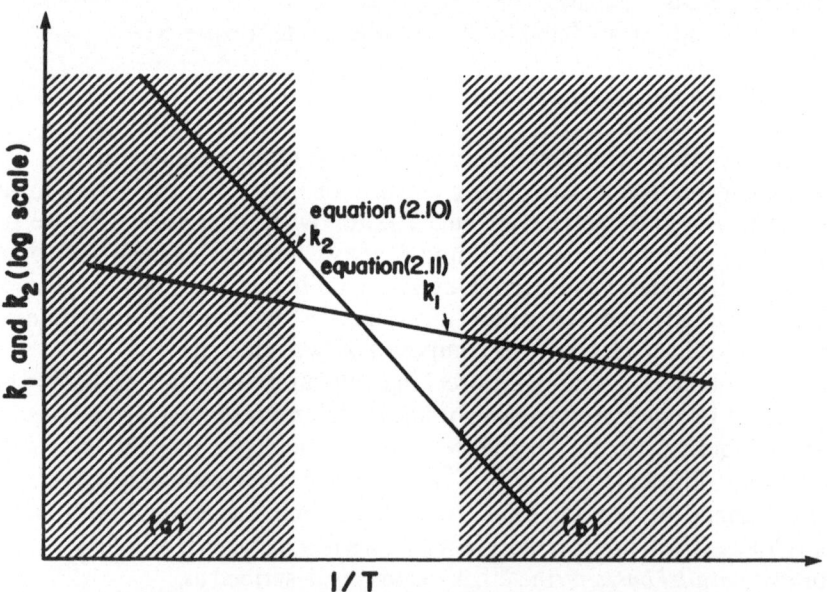

Figure 2.12. The schematic representation of the temperature dependence of the rate constants for two rate-controlling barriers.

and in zone (b) it is $\mathscr{k}_1$: hence in zone (a),*

$$\log v \simeq \log L_2 \mathscr{k}_2 = \log\left(L_2 \frac{kT}{h}\right) - \frac{\Delta G_2^+(W)}{2.3k}\frac{1}{T} \qquad (2.10)$$

while in zone (b),

$$\log v \simeq \log L_1 \mathscr{k}_1 = \log\left(L_1 \frac{kT}{h}\right) - \frac{\Delta G_1^+(W)}{2.3k}\frac{1}{T} \qquad (2.11)$$

and in the zone between,

$$\log v = \log(L_1 \mathscr{k}_1 + L_2 \mathscr{k}_2)$$

$$= \log \frac{kT}{h} + \log\left\{ L_1 \exp\left[-\frac{\Delta G_1^+(W)}{kT}\right] + L_2 \exp\left[-\frac{\Delta G_1^+(W)}{kT}\right]\right\}.$$

Accordingly, for mechanisms controlled only by breaking activations over two parallel barriers, the Arrhenius plot is composed of two straight lines in zones (a) and (b), and by a transition curve between the two, as shown in Figure 2.13.

When *two consecutive barriers* (with breaking activation only) control the

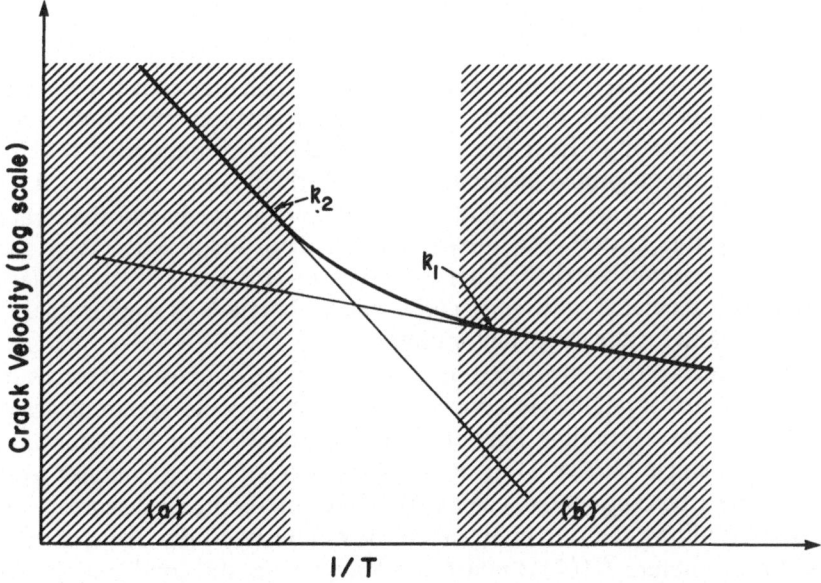

Figure 2.13. The Arrhenius plot of mechanisms controlled by two parallel energy barriers, with breaking activation only. The heavy line represents the measured behavior, while the light lines indicate the effects of the two individual rate constants.

---

* See footnote, p. 33 equation (2.7b).

velocity, the constitutive equation is

$$v = \frac{L_1 L_2 k_1 k_2}{L_1 k_1 + L_2 k_2} = \frac{1}{\dfrac{1}{L_1 k_1} + \dfrac{1}{L_2 k_2}}.$$

As an example, consider the double kink model with n spreading steps at the rate k_2 each. The constitutive equation is (see Appendix C)

$$v = \frac{1}{\dfrac{1}{L_1 k_1} + \dfrac{n}{L_2 k_2}}$$

and, therefore, in zone (a) where

$$\frac{1}{L_1 k_1} \ll \frac{n}{L_2 k_2}, \qquad v = \frac{L_2 k_2}{n};$$

in zone (b) where

$$\frac{1}{L_1 k_1} \gg \frac{n}{L_2 k_2}, \qquad v = L_1 k_1;$$

and in the zone between (a) and (b)

$$v = \frac{1}{\dfrac{1}{L_1 k_1} + \dfrac{n}{L_2 k_2}}.$$

Accordingly, as represented in Figure 2.14, the Arrhenius plot is composed of two straight lines in zones (a) and (b) and by the transition curve between. Note that the curvature is opposite to that of the parallel barrier kinetics.

Often, therefore, a simple inspection of the Arrhenius plot indicates the likely mechanism. From Figures 2.13 and 2.14 it follows immediately that when the crack velocity line curves *away* from the $1/T$ coordinate axis the process is controlled by *parallel* barriers; when it curves *toward* the $1/T$ coordinate, the mechanism is associated with *consecutive* energy barriers. A similar analysis by inspection is valid for the stress factor plots.

The Arrhenius plot is measured at constant microstructure and stress factor. The slope of each straight-line zone is proportional to the apparent activation energy of the corresponding barrier; from this ΔG^+ must be determined as shown for the single barrier. Derivation of the two activation energies from the slope and the pre-exponential values from the intercept, as discussed for the single barrier, allows the rate constants to be established fully. All characteristic values of the structure ΔG^+, α and L are known from the combined analysis of the stress factor and temperature dependence of the crack velocity.

The activation energy ΔG^+, the work constant α and the step size L, being microstructural characteristics, must always be measured. Once this is done, they are compared with the corresponding quantities calculated on the basis of the theory of the assumed mechanism. The constitutive equation cannot be

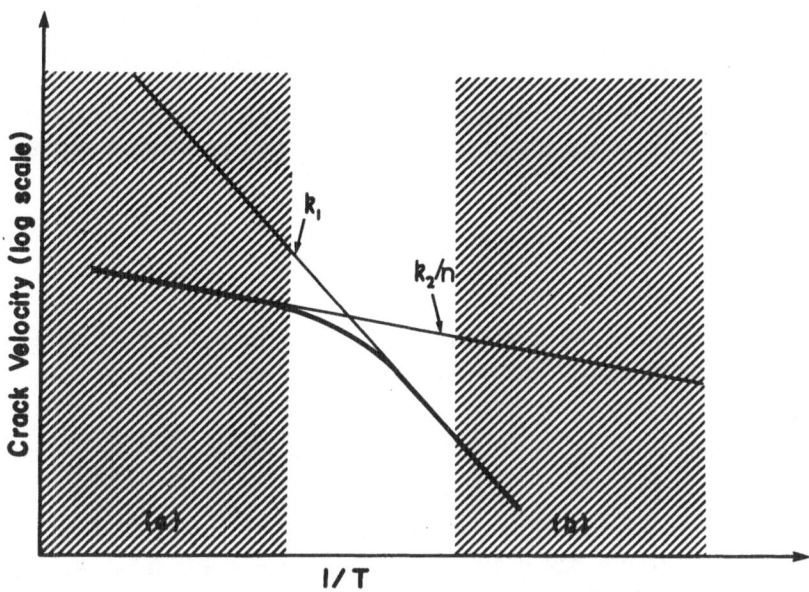

Figure 2.14. The Arrhenius plot of the mechanism with two consecutive barriers with breaking activation only.

considered valid unless the measured and calculated theoretical values are in agreement [28, 30, 31].

The work, *W*, is expressed through fracture mechanics and kinetics concepts. Because the fundamental and most developed branch of fracture mechanics describes the conditions of crack extension in linearly elastic solids, the stress intensity factor *K* or the crack-extension force $\mathcal{G}$ was used here as the definition of the work that acts in the crack-tip zone. However, the kinetics analysis is equally valid for elastic—plastic, plastic, and time-dependent plastic conditions; for these the appropriate fracture mechanics parameters, that define the work, should be used.

For expediency, then, this discussion is developed for conditions where the work in the crack-tip zone is linearly dependent on the stress factor (specifically *K* or $\mathcal{G}$). While most crack growth mechanisms are indeed well represented by this simple form, the extension of the analysis to types of work that are not linearly dependent presents no conceptual difficulties, and follows the procedure described above. First, however, the existence of non-linearly stress dependent mechanism must be established. This is a difficult task, and discussion of it could lead beyond the scope of this book. It is simply noted here that a theoretical understanding, aided with a combined driving force—temperature change test and analysis system must be employed; this provides a powerful tool for the determination of the functional form of the work. These concepts follow the firmly established principles of statistical mechanics and the corresponding practices associated with thermally activated processes [1, 21, 23—27, 32, 33]: some of them will be elaborated in the following chapters [39, 40].

Kinetics with backward activation

As discussed before, it follows from the principles of statistical mechanics that at the atomic level both breaking and healing steps always occur, and that the healing effect can be disregarded only when the rate of backward activations is much smaller than the breaking rate.

When the healing rate is not negligible, the crack propagation rate associated with a single-energy-barrier mechanism is

$$v = L_b k_b - L_h k_h.$$

The corresponding velocity versus stress factor relation is shown in Figure 2.15.

Usually, in the high stress factor and low temperature range,

$$k_b = \frac{kT}{h} \exp\left(-\frac{\Delta G_b^* - W_b}{kT}\right) \gg k_h = \frac{kT}{h} \exp\left(-\frac{\Delta G_h^* + W_h}{kT}\right),$$

and the relation is expressed by a straight line, as in Figure 2.15. Evaluation of the experimental results can begin here: the $L_b(kT/h) \exp(-\Delta G_b^*/kT)$ and the α_b values are readily determined from the intercept and slope, respectively. Once the rate constant k_b is known, the healing rate constant can be calculated by rearranging the constitutive equation to the form

$$L_b k_b - v = L_h k_h$$

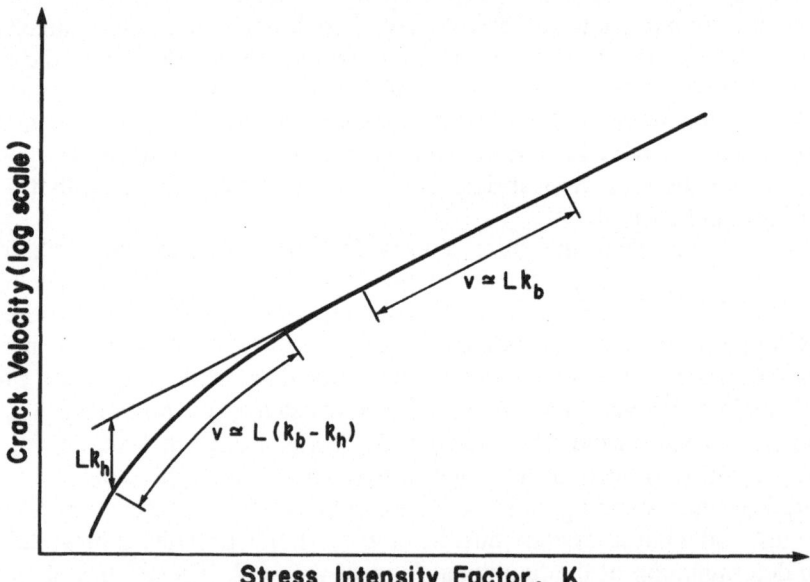

Figure 2.15. Schematic representation of the crack velocity behavior associated with a mechanism controlled by a single energy barrier with breaking and healing steps.

and replotting this equation in the $\log(L_b k_b - v)$ versus stress factor coordinate system. If indeed a single-barrier mechanism is in control, the plot results in a straight line.

When, however, the behavior is represented by a curve, the operation of a more complex process must be assumed. The analysis then proceeds by positing a two-barrier system that considers backward activation as well.

It has been shown that double-barrier kinetics with forward and backward activation is described for a parallel process as

$$v = L_{b1} k_{b1} - L_{h1} k_{h1} + L_{b2} k_{b2} - L_{h2} k_{h2}, \tag{2.12}$$

and for a consecutive process as

$$v = \frac{1 - \dfrac{L_{h1} L_{h2}}{L_{b1} L_{b2}} \dfrac{k_{h1} k_{h2}}{k_{b1} k_{b2}}}{\dfrac{1}{L_{b1} k_{b1}} + \dfrac{1}{L_{b2} k_{b2}} + \dfrac{L_{h1} k_{h1}}{L_{b1} k_{b1} L_{b2} k_{b2}} + \dfrac{L_{h2} k_{h2}}{L_{b1} k_{b1} L_{b2} k_{b2}}}. \tag{2.13}$$

The formulation of the constitutive equation expressed by equations (2.12) and (2.13) requires further elaboration: this is given in Appendix C.

It is appropriate to stress again the difficulties involved in the full kinetics analysis of two-barrier mechanisms. However, the generally applicable, step-by-step procedure outlined above greatly simplifies the task. This is crucial when dealing with the problem of deriving quantitative descriptions of crack growth behavior, based on rigorous constitutive laws.

2.3. Comments and summary

The foregoing discussion showed how the limitations of empirical and semi-empirical relations can be overcome by employing the theoretically rigorous, systematic and generally applicable method of fracture kinetics.

Essentially, fracture kinetics is the method of establishing (1) the rate of each of the component steps of a crack growth process; (2) the quantitative effects of the associated loading, geometrical boundary conditions, thermal and chemical environment, and material characteristics; and, most importantly (3) the complete description of the crack growth velocity — that is, the constitutive equation of time-dependent fracture,

$$v = f(\text{load, geometry, thermal and chemical environment, microstructure}).$$

For most engineering applications it is satisfactory to consider the kinetics combination to be one of the following:
(i) parallel kinetics (parallel energy barrier system), where the rate constants represent independent mechanisms;
(ii) consecutive kinetics (consecutive energy barrier system), where the rate constants represent dependent mechanisms; or

(iii) kinetics representing a complex mechanism consisting of consecutive systems combined in parallel.

The analysis involves describing the elementary rate constant for each step of a mechanism, and the combination of these steps. First, the simplest model, a single energy barrier with breaking activation only, is considered and determined fully; then, if necessary, models of increasingly greater complexity are considered until the constitutive equation of time-dependent fracture,

$v = f[\mathscr{k}$ (load, geometry, thermal and chemical environment, microstructure)]

is defined.

To do this, the rate of each constituent step is expressed by a function that is universally valid for all thermally activated processes: the elementary rate constant

$$\mathscr{k} = \frac{kT}{h} \exp\left[-\frac{\Delta G^+(W)}{kT}\right].$$

Tests are carried out to indicate the effects of the work W and the temperature T. Two similar diagrams, velocity versus stress factor and velocity versus reciprocal temperature, $1/T$, are then constructed, and loading and thermal conditions selected to allow isolation of each elementary rate constant. The corresponding rate constant is fully evaluated from the information obtained from the two diagrams. The rate constant is now expressed as

$$\mathscr{k} = \frac{kT}{h} \exp\left(-\frac{\Delta G^+ \mp \alpha K}{kT}\right).$$

As noted, fracture kinetics always adopts the simplest possible model (the simplest possible combination of rate constants) that provides a satisfactory explanation of the mechanism. In the analysis, rate constants are established through a sequence of steps. In each step, the constitutive equation is rearranged so that its left-hand side expresses both the measured velocity and the previously established rate constant expression. Temperature and load conditions are then selected so that the step with a dominant effect over a large enough segment of the velocity versus stress factor plot is isolated and the corresponding rate constant determined: that is, a dominant single rate constant appears on the right-hand side of the equation, the other possible rate constants being negligibly small in the selected load and temperature range. This sequence is repeated until all rate constants and rate constant combinations necessary to answer a specific research query, or design engineering task, are established.

The analysis of temperature dependence follows the pattern established for the Arrhenius diagrams, combined with the method for the stress factor analysis just described. On the logarithmic ordinate the quantity is plotted versus the reciprocal temperature on the linear abscissa, and the straight-line segment of the resulting curve can be evaluated directly to obtain the apparent activation energy. From this, the true activation energy, ΔG^+, reflecting the microstructure of the material, can be determined using the results of the stress factor analysis. Again, as in the stress factor analysis, one rate constant is evaluated, and joins the terms on the left side of the equation. The next step is carried out by replotting and evaluating the next rate constant.

Because each step in this method follows the same pattern, extension of the analysis to any number of rate constants presents no conceptual difficulties. The procedure provides a powerful, systematic framework for defining even the most complex of crack growth processes. By quantifying the individual rate constants and establishing their kinetics combination, and thus the associated barriers, the crack growth mechanism (including the quantitative definition of the microstructural characteristics of the material, $\alpha_i, \Delta G_i^+, L$) is fully expressed.

A fully rational constitutive equation must be derived through a complete analysis of stress factor and temperature effects, accompanied by careful consideration of service conditions as well as fractographic and, if necessary, chemical and metallographic studies. It is also essential to compare the conclusions derived from the analysis with the theoretical description of the mechanism. The processes of crack growth can be fully understood only after an investigation of this extent yields a quantitatively and conceptually satisfactory mathematical definition.

This description of the analysis concentrated on work conditions that are dependent on the linear stress intensity factor. Alternatives, such as crack driving force-dependent work factor, or other non-linear stress effects, can be considered using the same analytical techniques; some examples will be shown later.

Finally, it might be observed that in any branch of engineering, maturity is attained through the establishment of a scientific, rational foundation. Fracture kinetics meets this criterion. It is derived from statistical thermodynamics and is, therefore, the unique theory of time- and temperature-dependent crack growth processes and their corresponding constitutive equations. It provides a rigorous, physically based analytical method; it offers a full, quantitative description of crack growth behavior that is valuable for engineering applications; and it is the only dependable means for extrapolating outside the range of tested behavior. As well, its scientific, rational foundations make it a powerful instrument for research and development.

PART 2
DETERMINISTIC CONSTITUTIVE LAWS

As noted in Part 1, subcritical crack growth is controlled by thermal activation: this is perceived macroscopically as time- and temperature-dependent behavior that varies with the material, the loading and the environmental conditions.

When the sensitivity of the crack velocity to temperature or loading rate is negligibly small, a particular problem can be studied adequately as nonthermally activated — assuming, of course, that the approximation indeed simplifies the model. For example, this concept can be used in cases of mechanical, or reference, fatigue at temperature ranges where the process exhibits only minor sensitivity to cycling frequency. However, test results must be extrapolated with great care to avoid erroneous predictions. Particularly relevant examples are provided by corrosion fatigue, as will be shown in Section 2.6.

The rigorous anlaysis of time- and temperature-dependent crack growth is based on the principles of fracture kinetics. It was shown that the resultant constitutive laws describe the rearrangement and breaking of atomic bonds, observed macroscopically as subcritical crack growth, and that fracture kinetics incorporates two theories: rate theory and fracture mechanics.

Generally, rate theory defines the frequency of the elementary steps of thermally activated processes. In the case of crack growth, it defines the frequency of the steps causing the movement of the crack front, and describes them fully by the rate constant

$$\mathcal{k} = \frac{kT}{h} \exp\left(-\frac{\Delta G^* \mp W}{kT} \right). \tag{2.14}$$

Attention is drawn to the symbolism: the $\mp$ sign should not be considered as representing a symmetrical energy barrier, but rather as a shorthand notation expressing both breaking and healing activations. In equation (2.14) the effects of the microstructure of the material, and its chemical environment, are expressed quantitatively by ΔG^+ and W. Loading and geometrical conditions are defined rigorously by the work, W, as follows from fracture mechanics:

$$W \propto K, \text{ or } W \propto \mathcal{G}, \text{ or } W \propto J\text{-integral, or } W \propto C^*,$$

or others, according to the particular conditions under investigation.

Because crack growth usually involves more than a simple sequence of identical elementary steps, fracture kinetics must be applied in the majority of cases. The combinations of these steps can be classified into one of three categories:

(1) parallel combination of independent processes;
(2) consecutive combination of dependent processes;
(3) parallel combination of consecutive processes.

The kinetics combination of the elementary rate constants depends on the rate controlling mechanism and therefore expresses quantitatively the effects of material composition and structure; chemical and thermal environments; and loading and geometrical conditions. Because it represents rigorously all factors that control crack velocity, it provides a quantitative description for design and test engineering and for improving fracture control. Furthermore, the principles of fracture kinetics assure that the description is unique.

These concepts, developed in the previous chapters, will now be applied to the analysis of thermally activated crack growth processes, and to the establishment of their constitutive laws, for important practical engineering applications and for research. The following sections provide the descriptions of directly applicable constitutive equations; they also give numerical examples that will deepen and secure the understanding of the previous sections. From doing comes knowledge.

2.4. Environment-assisted cracking

When a component is exposed to a reactive environment, cracks often grow under sustained (static or varying) loads that could be carried safely in vacuum or in a non-reactive, inert environment.

Environmental effects can be induced by gas, liquid, or radiation surrounding the material; or they can be generated internally, as when hydrogen in the solid diffuses to the crack-tip zone. The environment-assisted cracking (EAC) process involves structural changes that weaken and break the atomic bonds in the matrix material of the crack-tip zone, as illustrated in Figure 2.16. This type of crack growth is typical of metallic, polymeric and ceramic materials.

The degrading action of the environment takes place in several steps. Transport of the environment to the crack-tip zone is followed by: attraction of the degrading material to the solid and its possible absorption; diffusion; changes in the chemical composition of the tip zone; and, in some cases, the transport of the reaction products into the environment. The chemical reactions are always thermally activated, while the transport processes may or may not be. Transport of the reactant material is also affected by mechanical and thermal convection, and both the transport and chemical processes are influenced by the geometry of the crack-tip zone. The exposure of fresh surfaces after each crack propagation step creates a renewed chemical environment. Electrochemical processes, and temperature and pressure gradients, have strong effects as well. For the internal degrading effect see Figure 2.16 and [1—18].

In this analysis, a relatively simple model is used to describe a very complex process. However, it is sufficiently complete to guide many sophisticated research and development studies, and practical engineering applications. In particular, it must be noted that the mechanisms of stress corrosion cracking (SCC) discussed below are associated with rather special cases. For extension of the analysis to the actual operating processes on hand, references [1—18] should be consulted.*

It is customary to represent the experimental results of stress corrosion cracking in the velocity versus stress intensity factor (or crack driving force) coordinate system, shown schematically in Figure 2.17. Regions I, II and III demonstrate typical SCC behavior; the figure also shows the threshold region, where the crack velocity approaches zero asymptotically at an applied stress that corresponds to K_{th}, often denoted as K_{ISCC}. At high stress intensities, the critical crack velocity K_c is reached; it is to be noted that the theory of thermally activated processes loses its validity well below the critical velocity

---

* References to Chapter 2, Part 2, are given on page 140.

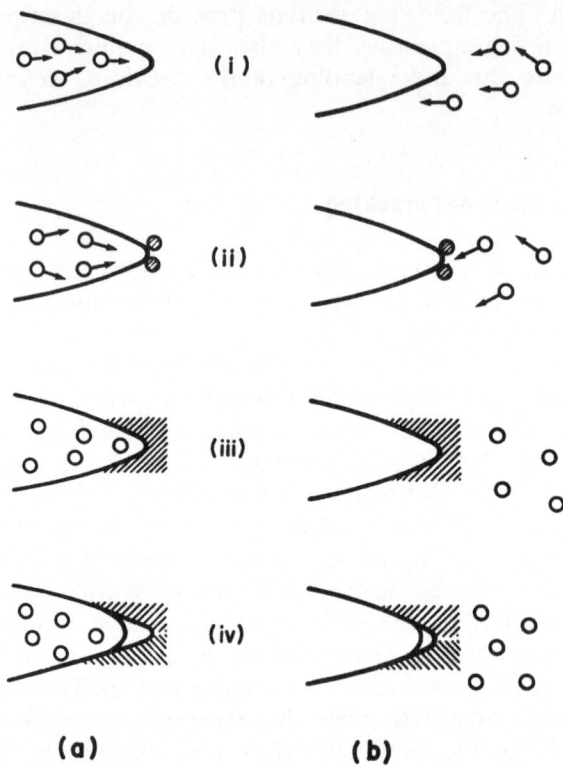

Figure 2.16. The transport of molecules and ions to the crack-tip zone, followed by the formation of a chemically altered zone, and crack propagation into the weakened region. (a) External environment; (b) internal environment. Figures (i) to (iv) illustrate in sequence the transport of the reactant to the crack tip, the chemical bonding of the reactant in the crack-tip zone, the development of a chemically altered region, and the crack growing into this zone. This sequence of environment-assisted crack growth repeats in cycles.

state. Nevertheless, the following discussion provides a valid analysis that includes all three Regions and the threshold zone as well.

First, consider Regions I and II in Figure 2.18. Because they are represented by a straight line in the semi-log coordinate system, each can be perceived as being controlled by a single rate constant, k_I and k_{II}, respectively. Next, the behavior represented in the figure suggests that Regions I and II compose a consecutive system. If this is true, then equation (2.13) is valid. Often, but not necessarily, the k_h terms are negligible because a single rate constant represents each region: accordingly, equation (2.13) becomes

$$v = \frac{L_I k_I L_{II} k_{II}}{L_I k_I + L_{II} k_{II}} = \frac{1}{\dfrac{1}{L_I k_I} + \dfrac{1}{L_{II} k_{II}}} \qquad (2.15)$$

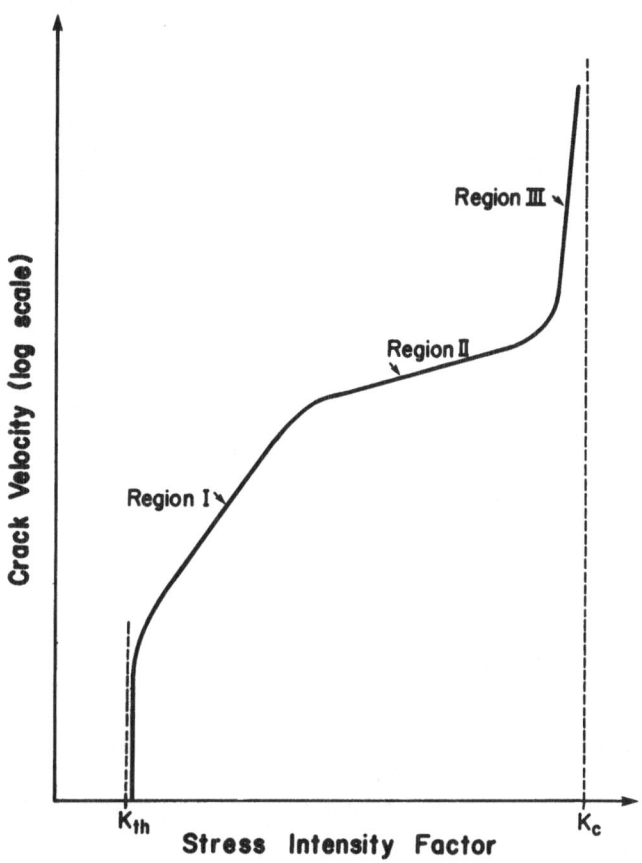

Figure 2.17. A schematic representation of the typical SCC process.

when the symbols I and II, appropriate to SCC, are used. L is the product of the crack growth that occurs with each activation and of the environment, expressing such effects as concentrations raised to their order of reaction, stoichiometry, hydrostatic and gas pressure, transport-related chemical and physical constants, molecular mass, etc., as the specific process requires. For explicit inclusion of these, see references [8, 13—16, 20] and Appendix C.

In Region I, the extension of the line associated with k_{II} is much above that of k_I; because $L_I k_I \ll L_{II} k_{II}$, equation (2.15) is reduced to the simple velocity relation $v = L_I k_I$, in agreement with the assumption that Regions I and II are sequential. In Region II the relative positions of the two straight lines representing k_I and k_{II} are reversed; by following the same argument, equation (2.15) reduces to $v = L_{II} k_{II}$.

Consider that in Region I the rate controlling mechanism is stress enhanced chemical bond breaking, as it is in glasses, while in Region II the crack extends into the zone weakened by transport controlled corrosion.

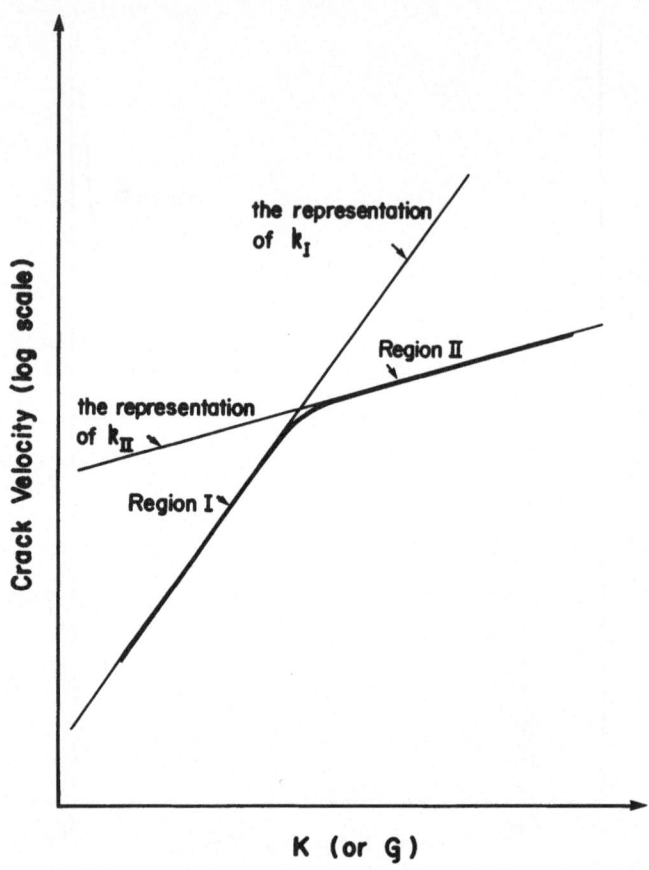

Figure 2.18. The schematic representation of Regions I and II. The heavy line indicates the observed behavior; the light lines represent the individual rate constants.

In Region I the stress factor, and thus the mechanical work, are small; there-fore, the bond-breaking rate is low. Indeed, Figure 2.18 shows that the transport of the degrading material occurs much more quickly than the actual breaking of the bond, and, therefore, the observed velocity is associated with chemical bond-breaking. Frequently, the crack velocity in Region II is almost independent of the stress factor. This indicates that k_{II} is a rate constant associated with transport, such as diffusion, processes that depend only slightly on mechanical effects. In this region the load, and thus the bond-breaking rate, are high; when the transport step is completed, the crack propagates so quickly that no time delay is observed on the macroscopic scale. Consequently, the propagation rate equals the transport rate, and the conclusion is reached again that k_{II} is a transport rate constant.

In the transition zone between Regions I and II where $L_1 k_I \simeq k_{II} L_{II}$ both chemical reaction and transport affect the macroscopic behavior. Accordingly, a gradual curving between Region I and Region II is observed, and the velocity is described by equation (2.15).

It is to be noted again that this identification of the mechanism in Regions I and II should not be considered as universally valid: EAC is a very complex process indeed. The kinetics analysis method described here is widely applicable, and its principles are generally valid. However, to identify the various mechanisms under specific microstructural and environmental conditions, the basic quantitative information of fracture kinetics must be supplemented by chemical and metallographic means [3, 4, 14—18].

The individual activation energies, ΔG^+, and the work function coefficients, α, must also be evaluated for the regions, and then compared to the theoretically expected values for the material—environment complex. Accurate identification of a mechanism is only assured when the values obtained from the kinetics analysis agree with the results of theoretical calculation of α and ΔG^+, or values obtained from other tests, for the specific condition.

The process described above is, to be sure, lengthy and difficult; but it yields a description of great value for development and design purposes. The fact that it is tedious is not the consequence of the kinetics theory and method but of the well recognized character of thermally activated processes in general. Nature is often like that.

Consider now the threshold region, which is at a stress intensity range where the crack velocity steeply diminishes to zero. It was noted in Chapter 1 on the physical processes of thermal activation that, by definition, no macroscopic change is observed when a system is in equilibrium; however, at the microscopic level an active regrouping of atoms continues. It was also shown that a crack may be in equilibrium even though stress is applied: here, the rate of bond breaking may be compensated by an equivalent rate of bond healing: $L_b k_b = L_h k_h$. Thus, to describe the process in the threshold zone, bond healing must be considered.

Figure 2.19 illustrates the typical macroscopic behavior in the threshold zone. Note that in the kinetics description of the consecutive process, zero crack velocity signifies that the numerator of equation (2.13) is zero:

$$(L)(k_{bI} k_{dII} - k_{hI} k_{rII}) = 0^*$$ (2.16)

As previously defined, the rate constant k_I represents chemically enhanced bond breaking, while k_{II} is associated with transport (diffusion). Accordingly, these are now identified as k_{bI} and k_{dII} respectively. The opposite processes are chemical bond healing in Region I, represented by k_{hI}, and the reversal of

---

* Notice that for clarity the rate factor L, the contribution of each thermally activated step to the crack velocity, is not shown explicitly here nor in the following, but only in a bracket as a *reminder* that each rate constant must be multiplied by the appropriate L-factor. The reader should find no difficulty to include it in actual applications.

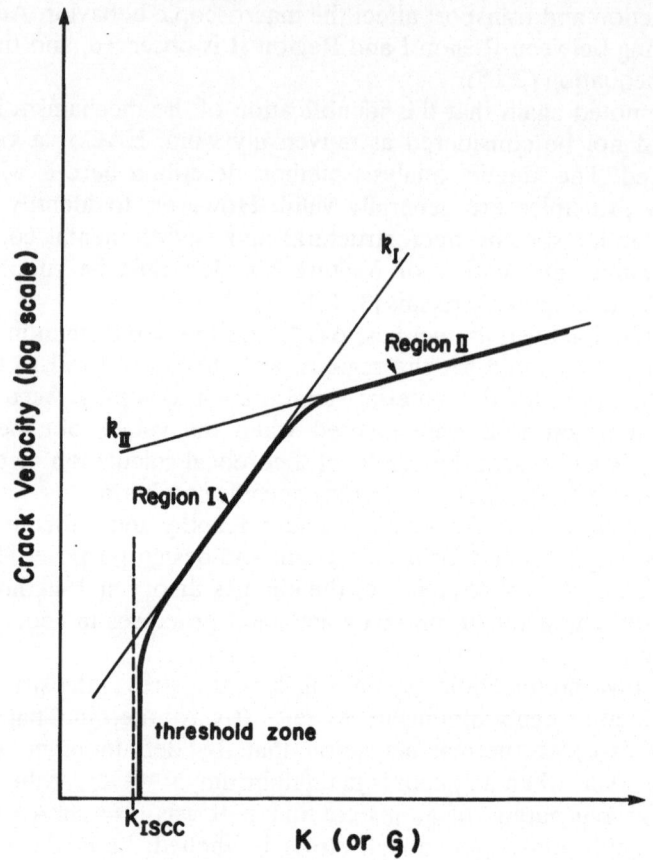

Figure 2.19. A schematic representation of behavior in the threshold zone. Compare with Figure 2.18 for the identification and role of k_I and k_{II}.

the transport step in Region II, represented by k_{rII}. The explicit expression of equation (2.16) is (see also Appendix C)

$$\frac{k_{bI}\,k_{dII}}{k_{hI}\,k_{rII}} = \frac{\exp\left[-\dfrac{\Delta G_{bI}^{+}(W) + \Delta G_{dII}^{+}(W)}{kT}\right]}{\exp\left[-\dfrac{\Delta G_{hI}^{+}(W) + \Delta G_{rII}^{+}(W)}{kT}\right]}$$

$$= \exp\left[-\frac{\Delta G_{bI}^{+}(W) + \Delta G_{dII}^{+}(W) - \Delta G_{hI}^{+}(W) - \Delta G_{rII}^{+}(W)}{kT}\right] = 1.$$

Consequently,

$$\Delta G_{bI}^{+}(W) + \Delta G_{dII}^{+}(W) = \Delta G_{hI}^{+}(W) + \Delta G_{rII}^{+}(W)$$

at K_{th}. It follows immediately that when the threshold behavior is controlled by the dynamic equilibrium at the atomic level, the threshold stress intensity is independent of the temperature. Figure 2.20 illustrates that this is indeed one of the conditions that have been observed [19, 20]; also in 18 Ni managing steel [21] and further substantiated by theory [16].

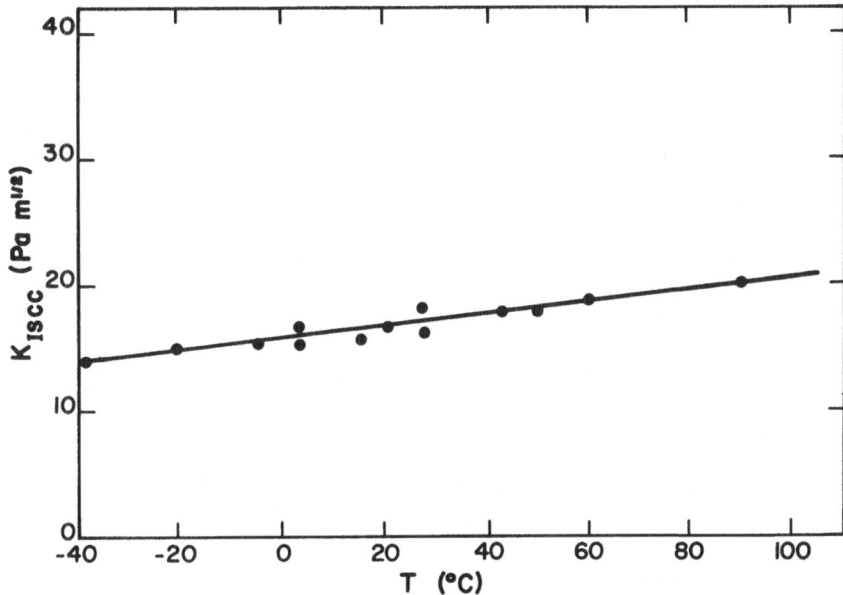

Figure 2.20. Illustration of the near-independence of the threshold stress intensity from the temperature, as measured in H-11 steel indicating a weak L effect [19, 20].

The threshold zone and Regions I and II are now fully expressed as

$$v = (L)\, \frac{k_{\text{bI}} k_{\text{dII}} - k_{\text{hI}} k_{\text{rII}}}{k_{\text{bI}} + k_{\text{dII}} + k_{\text{hI}} + k_{\text{rII}}},$$

with the interpretation of the rate constants and (L) as defined above.

The kinetics analysis then proceeds by determining the relation of Region II to Region III. Once more, as shown in Figure 2.21, the experimental results are represented by a straight line in the semi-log coordinate system of fracture mechanics; once more it is concluded immediately that crack propagation in this region is controlled by a single breaking activation rate constant, k_{bIII}.

How is then Region III associated with Region II? Consider again the parallel mechanisms with breaking rate constants only, which are written for these two regions as

$$v = L_2 k_{\text{b2}} + L_3 k_{\text{b3}} = L_{\text{II}} k_{\text{dII}} + L_{\text{III}} k_{\text{bIII}}; \qquad (2.17)$$

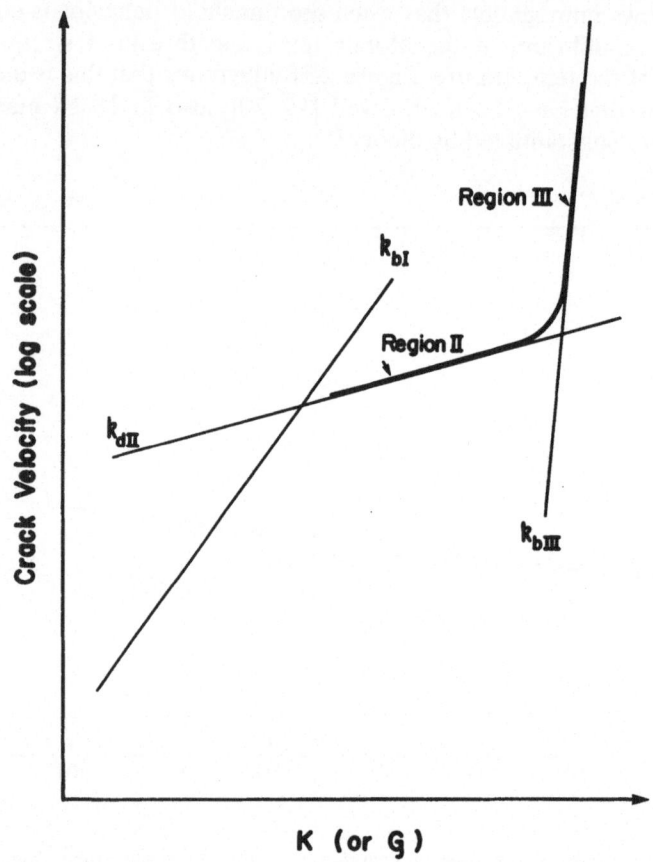

Figure 2.21. A schematic representation of the typical Region III behavior and its relation to Region II. To interpret the heavy and light lines, compare with Figure 2.18.

while consecutive mechanisms are expressed by equation (2.15) as

$$v = (L) \cfrac{1}{\cfrac{1}{k_{b2}} + \cfrac{1}{k_{b3}}} = (L) \cfrac{1}{\cfrac{1}{k_{dII}} + \cfrac{1}{k_{bIII}}} .$$

Figure 2.21 shows that $k_{dII} \ll k_{bIII}$. Clearly, the faster rate constant dominates the process in Region III, and the two mechanisms act in parallel: it is equation (2.17) that describes the velocity in the transition zone.

The physical process of Region II has been identified as controlled by the transport step because chemical bond breaking occurs more quickly. At the higher stress intensities of Region III, corrosion may have no role in bond breaking. Because the process does not depend on environmental factors, cracks can grow into uncorroded zones. Furthermore, since the rate of corrosion-controlled crack growth is so much slower than the uncorroded bond breaking, its effect in the parallel process is negligible at the macroscopic level.

In the transition zone between Regions II and III the rate constants are of approximately equal value: both control the crack velocity.

Fractographic analyses confirm these kinetics interpretations. In Regions I and II the fracture surfaces are corroded, while in Region III they are essentially corrosion-free.

The Region III rate constant k_{bIII} is combined in a parallel kinetics with Regions I and II. Accordingly, the full constitutive equation of environment-assisted crack growth is

$$v = (L) \frac{k_{bI} k_{dII} - k_{hI} k_{rII}}{k_{bI} + k_{dII} + k_{hI} + k_{rII}} + L_{III} k_{bIII}.$$

It was pointed out in Part I that as a general principle, kinetics analysis employs the simplest model that answers adequately all questions raised in a particular inquiry; the above procedure provides a case in point. In this analysis, threshold zone behavior was of interest and the reverse rate constants were therefore included. If, on the other hand, the enquiry had focussed only on behavior associated with higher stress factors, this step would have been omitted. Similarly, when there is a need to understand the transport processes and the chemical reactions associated with EAC, their descriptions can be developed in detail and incorporated in the constitutive equation [9—30].

EXAMPLE 2.1. Figure 2.22 illustrates the results of SCC experiments carried out on sapphire (Al_2O_3) to test the effect of water vapor concentration on crack velocity [8, 30, 65]. The results indicate that the kinetics analysis can be carried out satisfactorily using the constitutive equation for consecutive processes, without healing activation. The crack velocity is, as before,

$$v = (L) \frac{1}{\dfrac{1}{k_I} + \dfrac{1}{k_{II}}}.$$

In Region I, measurements reveal that only the k_I rate constant is effective. Thus k_I can be isolated;

$$v = L_I k_I = L_I \frac{kT}{h} \exp\left(-\frac{\Delta G_I^* - a_I \mathscr{G}}{kT}\right)$$

and hence,*

$$\log v = \log L_I \frac{kT}{h} - \frac{1}{2.3} \frac{\Delta G_I^* - a_I \mathscr{G}}{kT}.$$

The intercept of the line representing the Region I behavior at $\mathscr{G} = 0$ gives $\log(L_I(kT/h)) - (\Delta G_I^*/2.3kT)$, while the slope provides the work factor a_I.

In Region II outside the transition zone the crack velocity depends on the concentration of the environment, rather than the applied load; behavior is

---

* A reminder: see page 33, equation (2.7b) in Part 1.

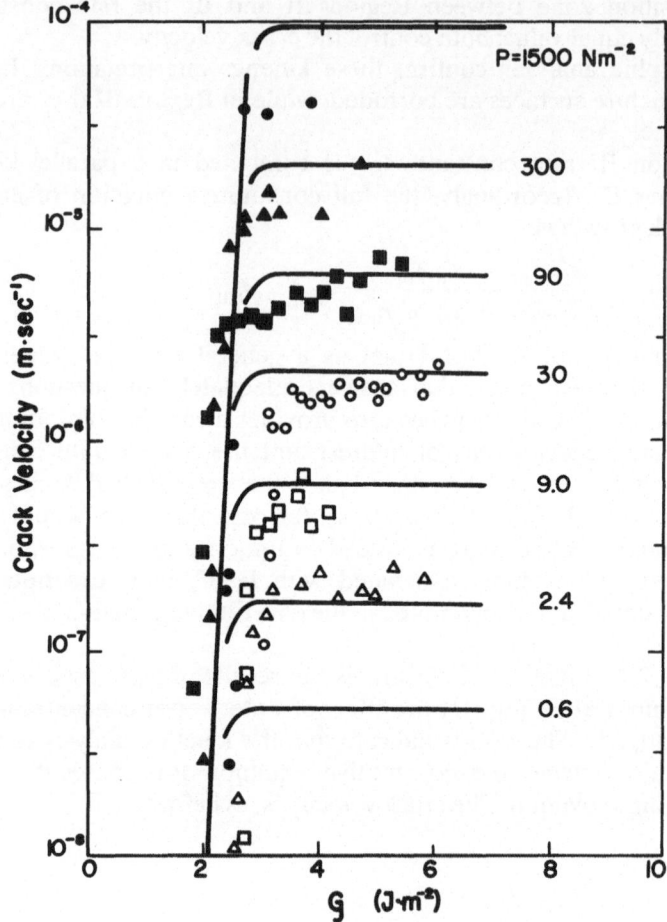

Figure 2.22. The crack growth velocity in stress corrosion cracking of sapphire exposed to water vapor at the indicated pressures [30, 65].

controlled by the rate constant k_{II}. Since the work has no effect, the crack velocity in Region II is

$$v = L_{II} k_{II} = L_{II} \frac{kT}{h} \exp\left(-\frac{\Delta G_{II}^{\ddagger}}{kT}\right),$$

or

$$\log v = \log L_{II} \frac{kT}{h} - \frac{1}{2.3} \frac{\Delta G_{II}^{\ddagger}}{kT},$$

where k_{II} is a function of the water vapor pressure

$$k_{II} = f(P_{vapor}).$$

Replotting the experimental results as $L(kT/h)(1/v)$ versus pressure gives the dependence of L on pressure. Measurements at various temperatures provide the complete information for the constitutive equation.

EXAMPLE 2.2. Figure 2.23 represents the stress corrosion cracking behavior

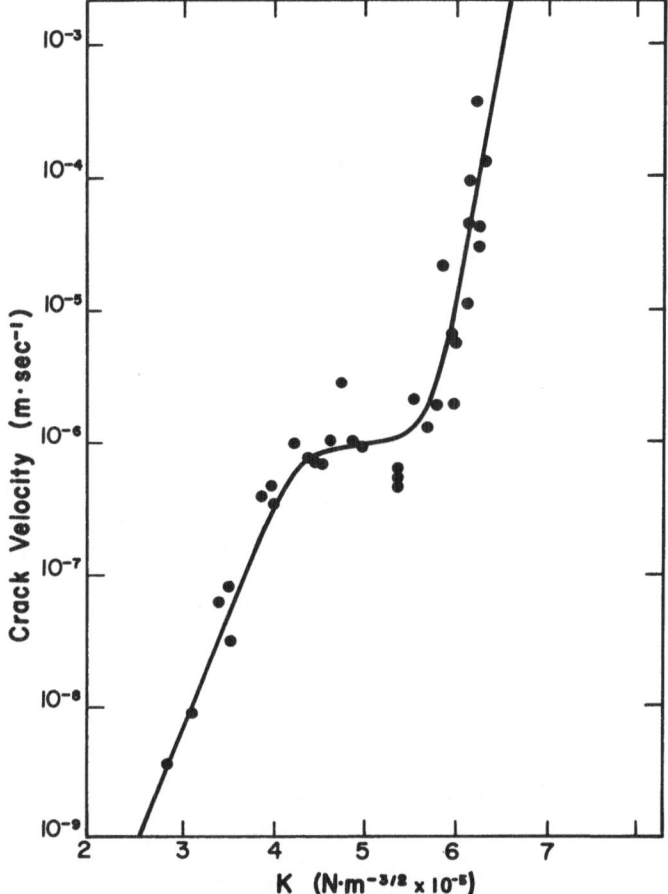

Figure 2.23. Stress corrosion cracking in glass, tested in octanol at 22 °C [19, 31].

of glass tested in octanol at room temperature [31]. All three regions were observed, but not the threshold zone. Accordingly, the constitutive equation is

$$v = (L) \frac{1}{\dfrac{1}{\mathscr{k}_{\mathrm{I}}} + \dfrac{1}{\mathscr{k}_{\mathrm{II}}}} + L_{\mathrm{III}} \mathscr{k}_{\mathrm{III}}.$$

In each region the straight-line behavior is sufficient to allow the approximation that only one rate constant controls the crack velocity. The individual rate constants can then be isolated easily.

In Region I

$$v = L_I k_I = L_I \frac{kT}{h} \exp\left(-\frac{\Delta G_I^* - \alpha_1 K}{kT}\right),$$

and in terms of the coordinate system

$$\log v = \log L_I \frac{kT}{h} - \frac{1}{2.3}\left(\frac{\Delta G_I^*}{kT} - \frac{\alpha_1}{kT} K\right).$$

Region II may be slightly dependent on the driving force. Between the two transition zones, crack propagation is controlled by the single rate constant k_{II}, because $L_I k_I \gg L_{II} k_{II}$. Hence, in this region*

$$v = L_{II} k_{II} = L_{II} \frac{kT}{h} \exp\left(-\frac{\Delta G_{II}^* - \alpha_{II} K}{kT}\right),$$

and

$$\log v = \log L_{II} \frac{kT}{h} - \frac{1}{2.3}\left(\frac{\Delta G_{II}^*}{kT} - \frac{\alpha_{II}}{kT} K\right).$$

The rate constant of Region III is isolated in a similar fashion.

The values, obtained from the intercepts on the velocity axis, and the α values derived from the slopes of the respective Regions, are given in Table 2.1.

Table 2.1. The parameters of Example 2.2.

Region	Intercept m s^{-1}	α J m$^{3/2}$ MN^{-1}
I	3.78×10^{-14}	1.6×10^{-19}
II	2.48×10^{-7}	1.1×10^{-20}
III	7.9×10^{-29}	3.56×10^{-19}

Temperature tests that provide ΔG^* are needed to evaluate the constitutive equation fully. Examples 2.4 and 2.5 illustrate the analysis of tests at various temperatures.

EXAMPLE 2.3. Figure 2.24 shows the results of crack velocity measurements for porcelain tested in water [32]. The example is similar to the previous one, but the test also included the threshold zone. Regions I and II are controlled by

---

* Recall that Regions I and II form a consecutive system.

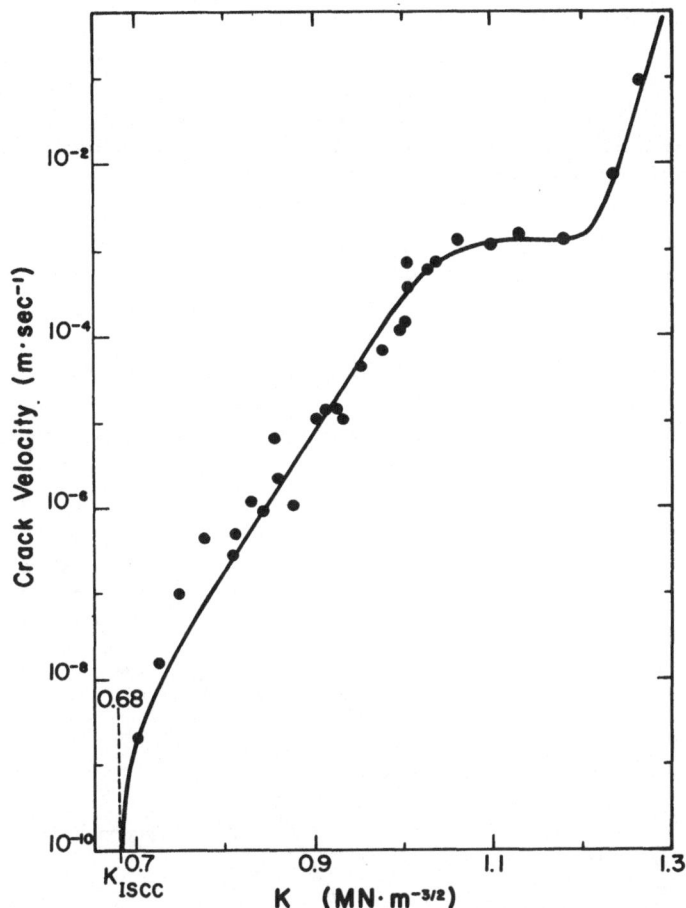

Figure 2.24. Stress corrosion crack velocity as a function of the stress intensity factor K for porcelain, tested in water [26, 32].

the individual rate constants k_{bI} and k_{dII} respectively. Their analysis follows the method shown previously; the intercepts and work factor values are given in Table 2.2.

Table 2.2. The parameters of Example 2.3.

Region	Intercept $m\,s^{-1}$	α $J\,m^{3/2}\,MN^{-1}$
I	4.6×10^{-20}	1.46×10^{-19}
II	1.4×10^{-3}	0
III	1×10^{-43}	3.04×10^{-19}

At the threshold the crack velocity is zero and thus

$$(L)(k_{bI}k_{dII} - k_{hI}k_{rII}) = 0. \tag{2.18}$$

It can be seen from equation (2.18) that K_{ISCC} is affected by both mechanisms of the consecutive system. It is also clear that the threshold stress intensity depends on the backward activation rates over the two energy barriers as much as it depends on forward activation. Equation (2.18) thus indicates an appropriate direction for the development of SCC-resistant materials.

The complete constitutive equation can be established as an exercise; the calculated parameters of Regions I, II and III are given in Table 2.2.

EXAMPLE 2.4. Soda-lime silicate glass was tested in the temperature range of 275 K to 363 K [7], as shown in Figure 2.25.

The resulting information can then be employed for constructing the Arrhen-

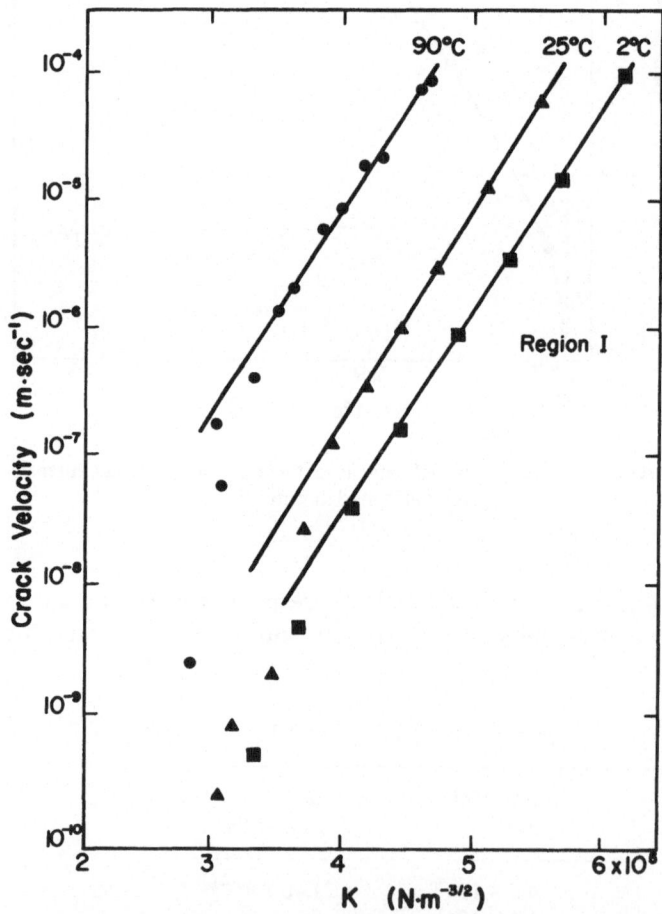

Figure 2.25. The measured crack propagation behavior of soda-lime silicate glass tested in water at three temperatures [7, 25].

ius plot of Figure 2.26. Evaluation indicates that the work function can be expected to be $\alpha_I = 0.6 \times 10^{-19}$ J m$^{3/2}$ MN^{-1}, and the activation energy $\Delta G_{bl}^+ = 1.37 \times 10^{-19}$ J; $L = 6 \times 10^{-11}$ m. Preliminary data gathered at only three temperatures are thus adequate to set the parameters for an experimental test program, including its extent, and time and cost allocations [25].

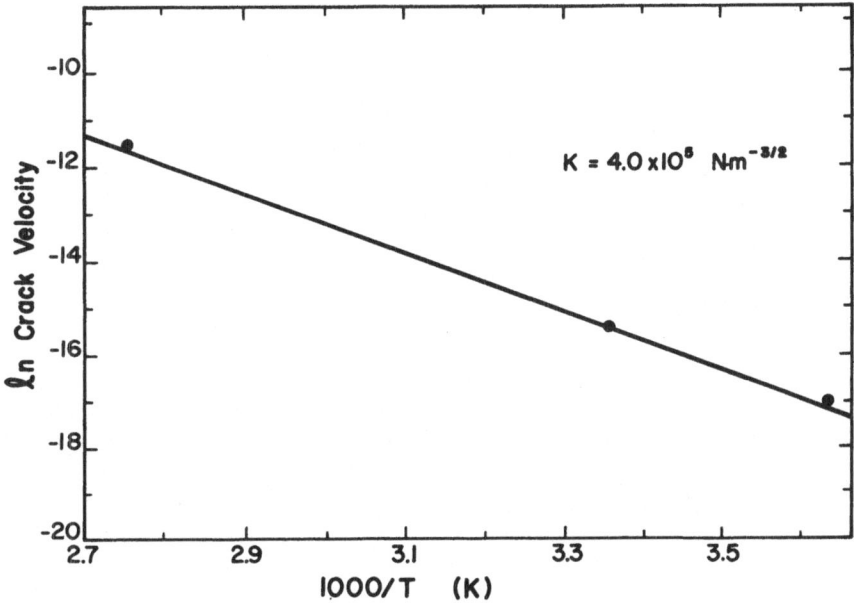

Figure 2.26. The Arrhenius plot constructed from the data represented in Figure 2.25.

A recent review of the fracture kinetics of SCC in glass includes a detailed description of the process [33].

EXAMPLE 2.5. Test results obtained in 61% lead glass over a range of 297 K to 625 K at seven temperatures is shown in Figure 2.27 [29]. The tests were carried out in vacuum; the results, therefore, represent the effect of the intrinsic lattice resistance, that is, thermally activated crack growth without corrosion.

Note that although the velocity is expressed as the function of K^2 rather than the more frequently used stress intensity factor K, the behavior shows the same linear relationship as in the log v versus K plots. This uncertainty of fracture mechanics in the identification of the relevant stress factor has been discussed in previous chapters.

Figure 2.28 shows the Arrhenius plot constructed from the data presented in Figure 2.27. Evaluation of the test data is left as an exercise; the results are given in [24].

Figure 2.29 shows that the test results support the constitutive equation. This indicates that it can be used with confidence to predict a crack growth

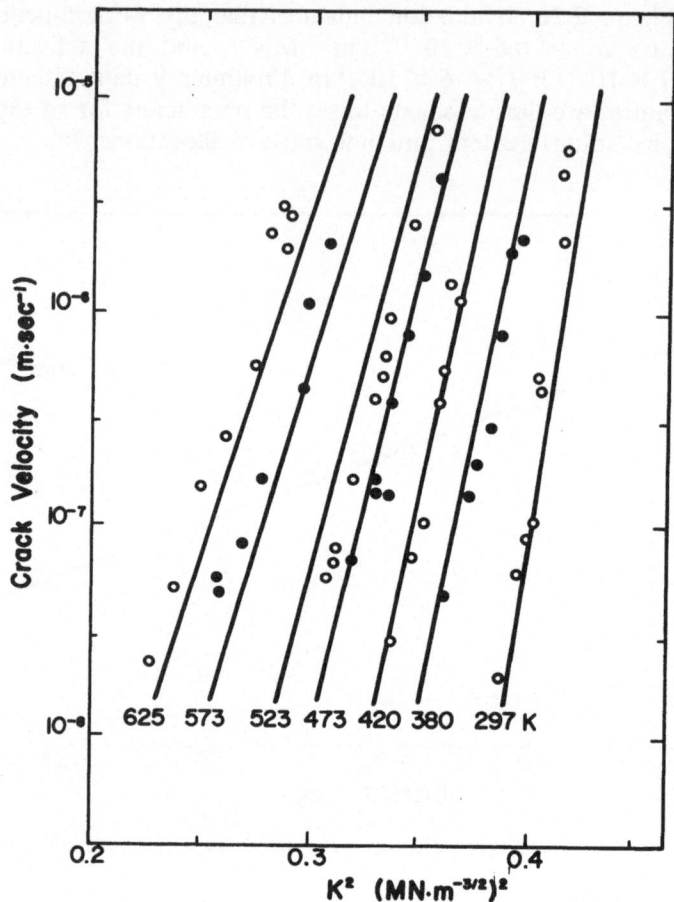

Figure 2.27. The measured crack growth velocity in 61% lead glass tested in vacuum [24, 29].

process associated with complex service conditions, as when temperatures vary over time, or when the crack moves across a region of temperature gradients.

2.5. Creep crack growth

Creep crack growth data are often expressed in terms of J^* or C^* because it has proven more useful in correlating measured crack behavior than the stress intensity factor; the representation of the velocity as a function of these often minimizes the scatter. The following discussion shows how fracture kinetics is applied to derive the constitutive equation of creep crack growth. The kinetics analysis follows essentially the same methods already demonstrated and will therefore be discussed only briefly.

Creep crack growth measured in Type 304 stainless steel at 925 K is shown in Figure 2.30. The stress intensity factor failed to describe crack velocity in this

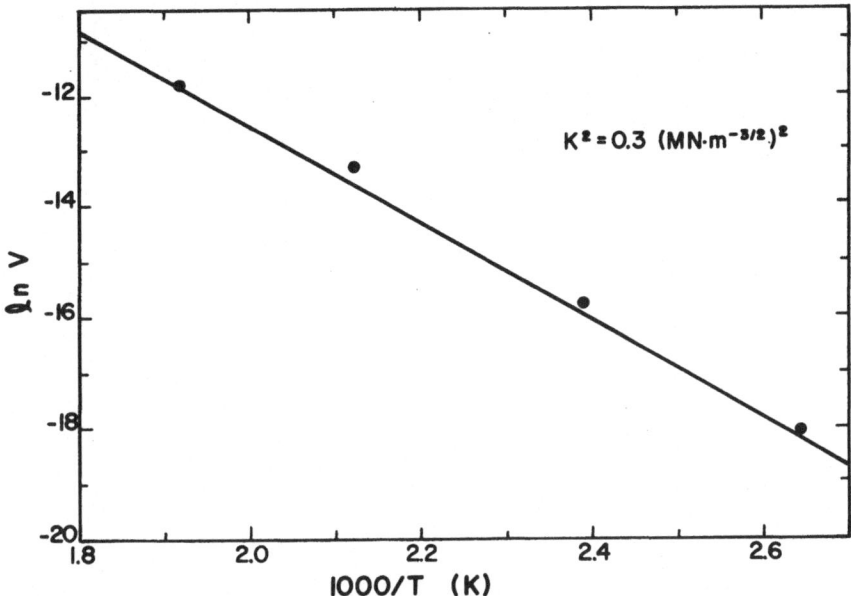

Figure 2.28. The Arrhenius plot constructed from the test data represented in Figure 2.27.

material at high test temperature, while J^* correlated the behavior satisfactorily [34, 35].

Often the empirical power function relation *

$$v = v_0 \left(\frac{C^*}{C_0^*} \right)^n$$

is used to describe the velocity. Here v_0 and C_0^* are scaling factors and, together with n, only empirical constants. Their evaluation is straightforward. From

$$\log v = \log v_0 + n \log C^* - \log C_0^{*n}$$

the intercept gives $\log v_0/C_0^{*n}$ and the slope the exponent n. Various mechanisms have been proposed to explain the power-function behavior: for these the literature can be consulted [3, 17, 18, 34—40].

Another empirical expression more closely related to thermal activation theory has also been proposed

$$v = v_0 \left(\frac{C^*}{C_0^*} \right)^n \exp \left(-\frac{\Delta E}{kT} \right),$$

where ΔE is an experimentally measured activation energy. This relation is also

---

* Note that this is a 'segmented' representation: a segment of the sinusoidal curve of Figure 2.31 is replaced by a straight-line approximation.

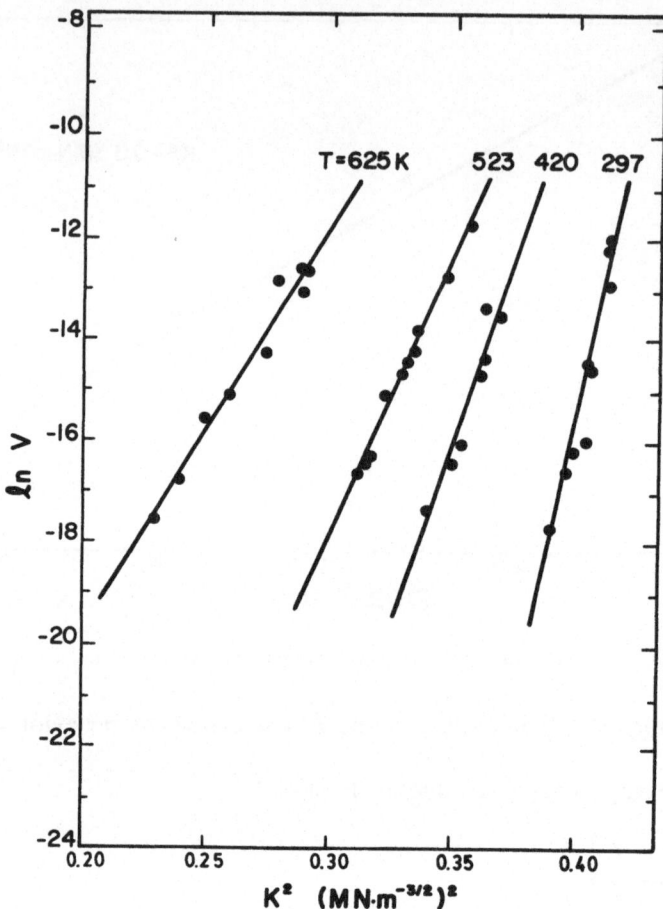

Figure 2.29. The figure illustrates the agreement between the ℓ_{bl} kinetics description and the test results.

empirical, or semi-empirical, and is thus associated advantageously with some understanding of the temperature effect. The following example demonstrates how kinetics theory can be applied to explain the empirical relations.

EXAMPLE 2.6. A simple kinetics description demonstrates the method used to explain the physical meaning of the parameters in the power function expression

$$v = v_0 \left(\frac{C^*}{C_0^*} \right)^n.$$

For illustration purposes, consider the single-energy-barrier kinetics with

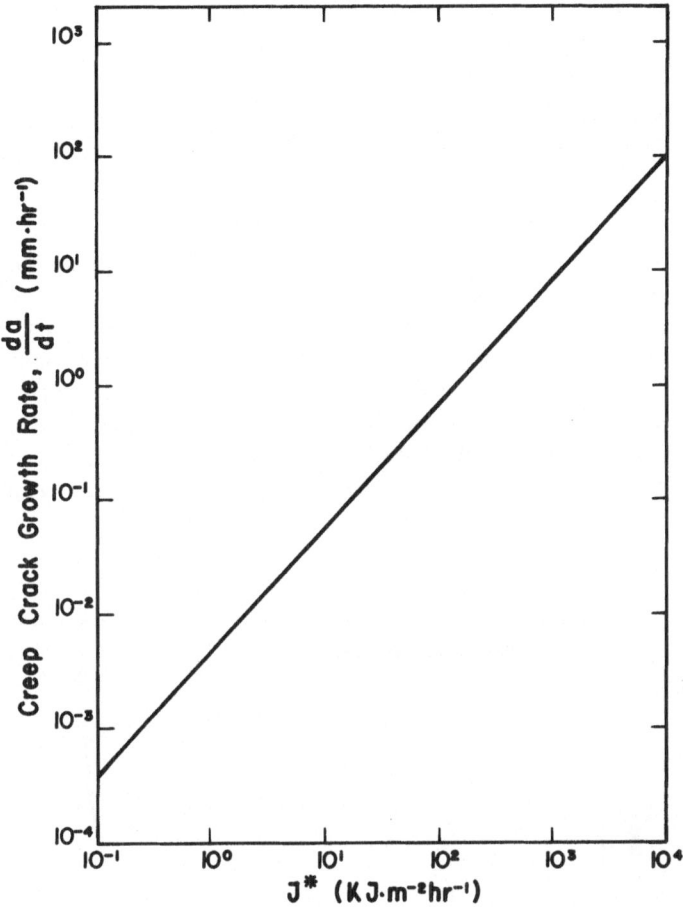

Figure 2.30. Creep crack growth measured in Type 304 stainless steel [34].

breaking and healing rate constants

$$v = L \frac{kT}{h} \left[\exp\left(-\frac{\Delta G_b^* - a_b C^*}{kT}\right) - \exp\left(-\frac{\Delta G_h^* + a_h C^*}{kT}\right) \right]$$

$$= v_0 \left(\frac{C^*}{C_0^*}\right)^n .$$

In the region where only the breaking rate constant is effective, the slope of the log v versus log C^* plot is defined as

$$\frac{\Delta \log v}{\Delta \log C^*} \cong \frac{d \ln v}{dC^*} \frac{dC^*}{d \ln C^*} = C^* \frac{d \ln v}{dC^*} ,$$

and is expressed by the kinetics theory using

$$\ln v = \ln L \, \frac{kT}{h} - \frac{1}{kT} \, (\Delta G_b^+ - a_b C^*)$$

as

$$\text{slope} = C^* \, \frac{a_b}{kT} \, .$$

The dependence of the stress exponent n on C^*, the stress factor, and on the temperature in the high-velocity region where the breaking rate constant controls the velocity, is well substantiated by many measurements. Figure 2.31 illustrates that crack growth behavior represented in a log velocity versus log stress factor coordinate system is sigmoidal: in the low stress factor region the crack velocity must go asymptotically to the threshold value, be it zero or not; in the high-velocity region it must approach the critical stress factor.

The slope of the sigmoidal curve is indeed a function of the stress factor; the

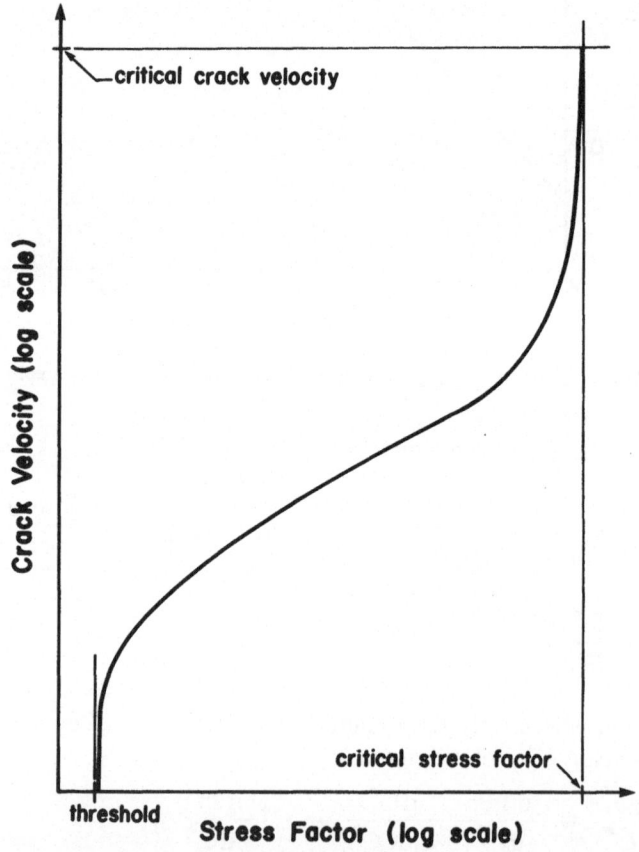

Figure 2.31. Schematic representation of a typical crack-propagation behavior.

constant stress exponent condition provides a good approximation only in the region of the inflexion point.

When the healing rate constant is also effective, the kinetics is expressed as a complex function of stress factor and temperature:

$$\text{slope} = C^* \frac{d \ln v}{dC^*}$$

$$= C^* \frac{a_b}{kT} \frac{1 + \dfrac{a_h}{a_b} \exp\left(\dfrac{\Delta G_b^+ - \Delta G_h^+}{kT}\right) \exp\left(-\dfrac{a_b + a_h}{kT} C^*\right)}{1 - \exp\left(\dfrac{\Delta G_b^+ - \Delta G_h^+}{kT}\right) \exp\left(-\dfrac{a_b + a_h}{kT} C^*\right)}. \tag{2.19}$$

At high temperature, the kinetics tend to lead to the symmetrical energy barrier condition

$$\Delta G_b^+ = \Delta G_h^+ = \Delta G^+ \quad \text{and} \quad a_b = a_h = a,$$

in which case

$$v = 2L \frac{kT}{h} \exp\left(-\frac{\Delta G^+}{kT}\right) \sinh\left(\frac{aC^*}{kT}\right).$$

For example, this relation is found in the diffusion-controlled growth of cavities, a mechanism very much associated with creep-fracture processes [36, 38]. For this mechanism, equation (2.19) reduces to

$$\text{slope} = C^* \frac{a}{kT} \coth \frac{aC^*}{kT}.$$

The corresponding temperature dependence of the reduced exponent n/C^* is represented in Figure 2.32.

The examples of Section 2.4 can be used to formulate other kinetics relations for the power-function equations. For example, they give the necessary guidance for the use and interpretation of the physical meaning of the power-function representation of EAC as well. These expressions often have only a purely empirical basis; the parameters are stress and temperature dependent, and the microstructural properties of the parameters are not readily identified. Consequently, they are not as powerful as the physically based, theoretically rigorous kinetics constitutive equations. When the limitations of power-function relations are clearly recognized and considered in applications, they provide a practical and simple alternative: however, extrapolations based on these expressions may greatly overestimate or underestimate crack propagation behavior, resulting in uneconomical or even dangerous conclusions. Here, interpretation through the physically based kinetics theory is of obvious importance.

Extensive studies of thermally activated crack growth processes under environmental effects and in creep conditions were reported recently [34, 41, 60, 63].

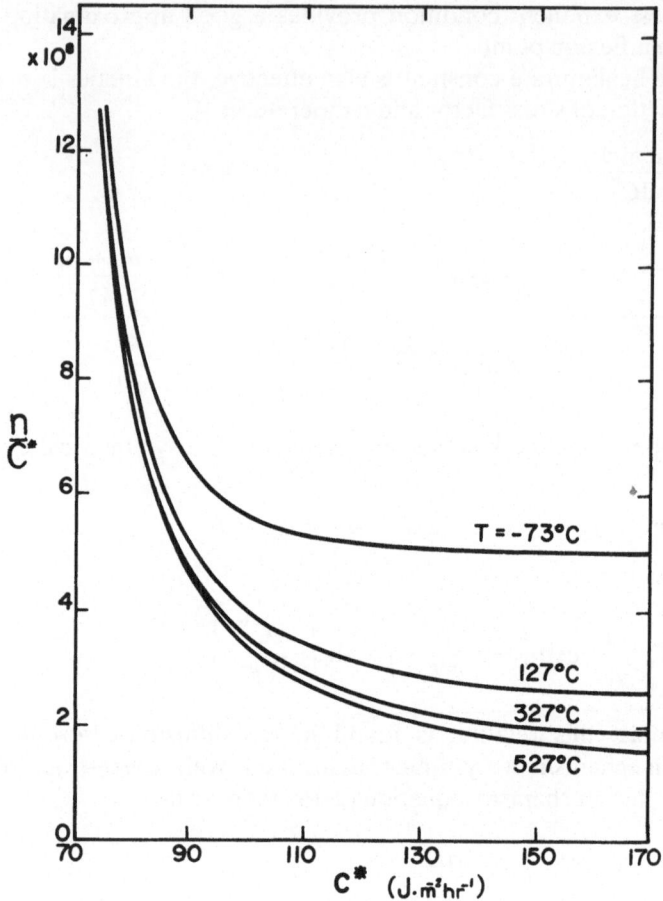

Figure 2.32. The dependence of the reduced exponent n/C^* on the temperature and the driving force.

Both sustained and fatigue loading conditions were considered. It was empha-sized that "both creep and environmental interactions are generally thermally activated processes and therefore are sensitive to changes in temperature".

2.6. Corrosion fatigue

There are strong indications that environment-assisted fatigue is the most frequent cause of material failure, and that this process involves the interaction of mechanical, geometrical, environmental and metallurgical variables.

It was proposed that the environmental fatigue crack rate, $(da/dN)_e$, is obtained by summing the appropriate crack growth rate expressions. The three basic types of fracture behavior are shown schematically in the rate versus ΔK (or K_{max}) coordinate system of Figure 2.33 [43—46]. Curve A, representing fatigue in vacuum or in a harmless, inert environment, is usually described by

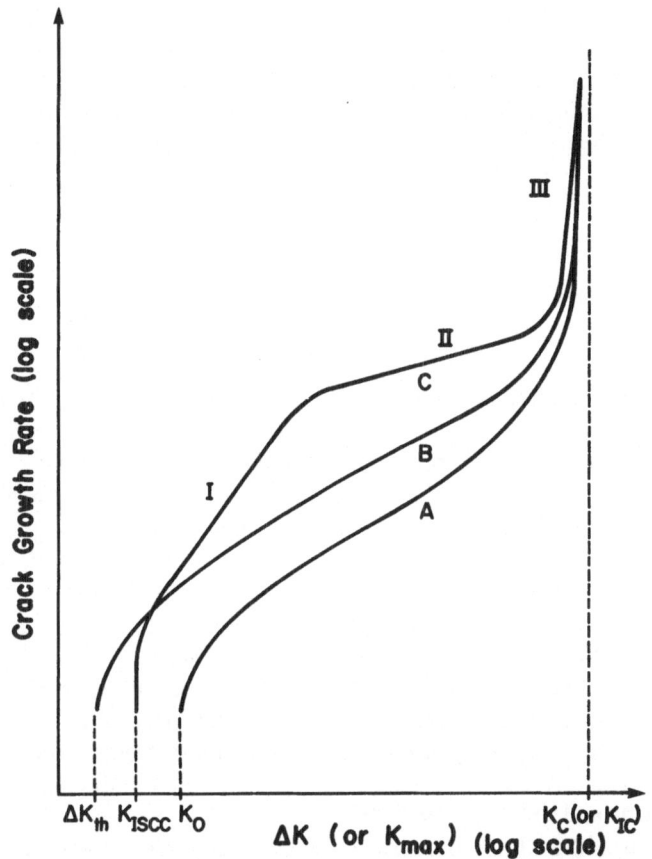

Figure 2.33. The schematic representation of the three basic types of crack propagation behavior. (A) Inert environment, the 'reference' fatigue; (B) synergistic mechanical and environmental fatigue; (C) the three regions of stress corrosion cracking. K_0, ΔK_{th}, and K_{ISCC} are the respective threshold values and K_c (or K_{IC}) is the critical stress intensity factor.

the Paris power-function relation as

$$\left(\frac{da}{dN} \right)_r = C(\Delta K)^n \tag{2.20}$$

where the subscript 'r' identifies the process as mechanical or reference fatigue; C is an experimentally determined constant; and n is also an experimental constant that often takes the value of 2. Equation (2.20) is valid in the linear range of curve A, for medium values of ΔK. The full description of the complete curve between K_0, the reference fatigue threshold stress intensity factor, and K_c, the critical stress intensity factor, may take the form

$$\left(\frac{da}{dN} \right)_r = \frac{C(\Delta K - K_0)^n}{(1 - R)K_c - \Delta K} , \tag{2.21}$$

where R is the ratio K_{min}/K_{max}. Equation (2.21) is one of the very many empirical power-function relations that have been formulated to describe specific sets of experimental conditions [1, 3, 4, 14, 18, 40, 46—53].

Curve B represents fatigue behavior influenced by both mechanical and environmental effects. Since the environment-assisted processes are the function of time and temperature, the representation of the crack propagation rate is often simplified to

$$\left(\frac{da}{dN}\right)_{cf} \propto (\Delta K)^n \exp\left(-\frac{\Delta G^+}{kT}\right), \tag{2.22}$$

where 'cf' denotes corrosion fatigue. Equation (2.22) is an empirical Arrhenius relation.

Curve C in Figure 2.33 represents a typical fatigue crack growth behavior associated with stress corrosion. Regions I, II and III are usually present; each is associated with a different mechanism. In some cases, crack propagation is described by power-function relations similar to those of equation (2.20). As has been shown, however, by the application of fracture kinetics, the behavior over the threshold zone and the three regions is described rigorously by physically based exponential relations [8, 14—16, 19].

The following discussion shows how fracture kinetics is applied to derive physically based constitutive relations that describe the rate of crack growth in environment-assisted cracking under fluctuating load. Some concepts of EAC are based on measurements carried out in sustained loading; load fluctuation introduces additional mechanisms [54].

When the loading fluctuates, reversal of the cycle causes crack closure: a large, elastically deformed body surrounds a relatively small, plastically deformed zone just ahead of the crack tip and the released energy closes the crack to the extent that actual rebonding may occur. The residual stress acts in rebonding as the applied stress does in bond breaking; it promotes the process. The rebonding rate constant k_{rb} is therefore

$$k_{rb} = \frac{kT}{h} \exp\left[-\frac{\Delta G_{rb}^+ - W_{rb}(K)}{kT}\right], \tag{2.23}$$

where ΔG_{rb}^+ and W_{rb} are the appropriate bond energy and work terms. The maximum applied load promotes the rebonding process and thus reduces the apparent activation energy $\Delta G^+(W)$ [46]. It is to be noted that here K is associated with the effective stress.

Although crack healing by backward activation and rebonding by crack closure are both thermally activated, the two processes differ, basically in that crack healing takes place against the applied stress W_b, while in rebonding k_{rb} is promoted by W_{rb}. The full description of the crack growth rate must include both k_b and k_{rb}, and the overall rate for any one region in Figure 2.33 takes the form

$$\frac{\mathrm{d}a}{\mathrm{d}N} \propto \ell_\mathrm{b} - (\ell_\mathrm{h} + \ell_\mathrm{rb}),\tag{2.24}$$

where the subscripts 'b', 'h' and 'rb' represent crack growth (breaking activation), crack healing (backward activation) and rebonding, respectively.

The relative influence of each of the three processes, and therefore of their individual rate constants, on the overall rate depends on the imposed conditions. As ΔK approaches the critical stress intensity, the backward activation rate falls to zero, while at lower ΔK values, particularly at the threshold zone, ℓ_h is also effective.

The effect of rebonding on crack growth time is shown in Figure 2.34. Growth begins only when the surfaces lose contact, and stops when the stress

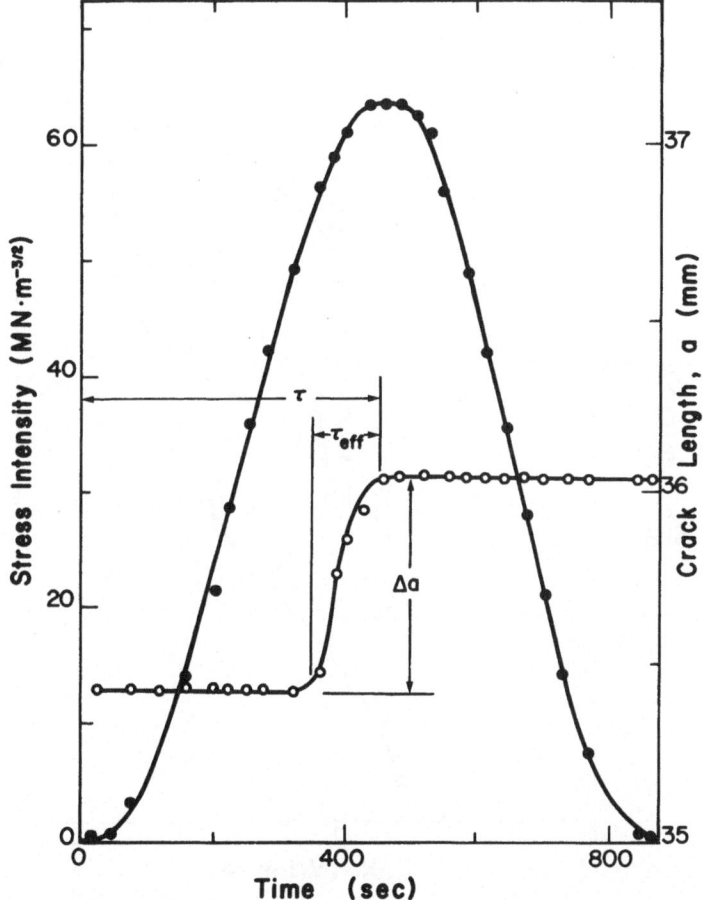

Figure 2.34. The relationship of the stress intensity and the growth of a fatigue crack during a cycle in a 12% chromium steel exposed to water at room temperature. The crack opens up and propagates during only part of the loading time. ● = stress intensity; ○ = crack length [46, 55].

intensity just passes its peak value. Thus, the time available for crack growth can be considerably less than the loading period of a cycle [55].

The crack growth rate for a specific waveform and frequency is expressed as a function of ΔK. With increasing frequencies, the crack growth rate, da/dN, decreases proportionally with the decrease of the effective time that is available for a cycle. As illustrated in Figure 2.35, this proportionality can be one-to-one: a ten-fold frequency increase results in a ten-fold growth rate decrease at the intermediate values of ΔK, in Region II. It follows from the kinetics theory that rebonding can change this ratio considerably from

$$\frac{(da/dN)_1}{(da/dN)_2} = \frac{\nu_2}{\nu_1}$$

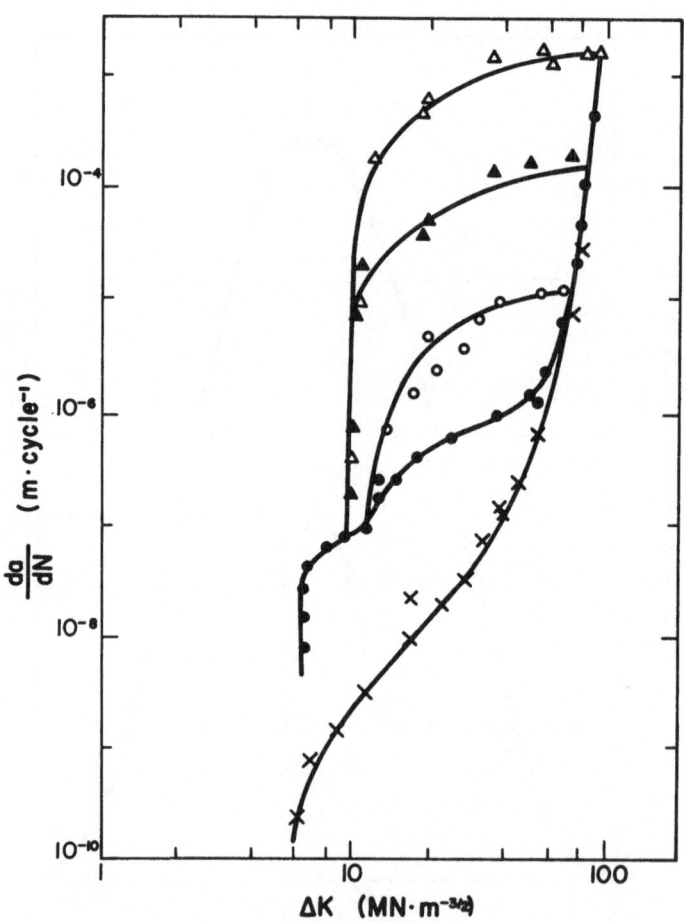

Figure 2.35. The effect of ΔK and frequency on the rates of fatigue crack growth in 4340 M steel exposed to water. Only some of the experimental points are shown. $\times$ = vacuum, the reference curve; $\bullet$ = 1 Hz; $\circ$ = 0.1 Hz; $\blacktriangle$ = 0.01 Hz; $\triangle$ = 0.001 Hz [46, 55].

to the expression

$$\frac{(da/dN)_1}{(da/dN)_2} \neq \frac{\nu_2}{\nu_1},$$ (2.25)

where ν is the frequency.

Figure 2.36 illustrates [56] that the modification of the rate/cycle frequency ratio indeed takes the direction predicted by equation (2.25).

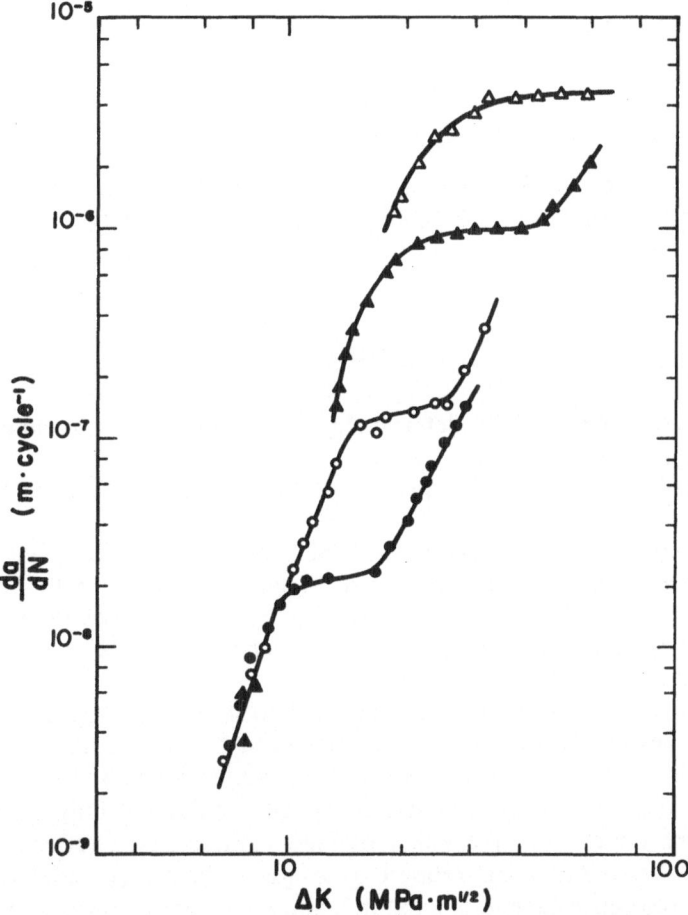

Figure 2.36. The effect of ΔK and frequency on the rate of fatigue crack growth in steel, exposed to 3.5% salt water. $\bullet = 10$ Hz; $\circ = 1$ Hz; $\blacktriangle = 0.1$ Hz; $\triangle = 0.01$ Hz [46, 56].

Studies of the influence of grain size, ageing conditions and environment on the crack closure process and on rebonding demonstrate clearly that microstructural conditions can be as important as mechanical and thermal factors. Microstructural considerations can be expressed by the kinetics equation of the fracture process and by the activation energy ΔG^+ and the work parameter α, as discussed.

In reference fatigue, if there is no backward activation: $\mathscr{k}_h = 0$, and the fracture kinetics expression of equation (2.24) reduces to

$$\left(\frac{da}{dN}\right)_r = \delta_b \mathscr{k}'_b - \delta_{rb} \mathscr{k}'_{rb}, \qquad (2.26)$$

where the semi-empirical — or empirical — rate constants are expressed as

$$\mathscr{k}'_b = A_b \exp(-G_b + w_b)$$

and

$$\mathscr{k}'_{rb} = A_{rb} \exp(-G_{rb} + w_{rb})$$

in formal agreement with equations (2.14) and (2.23), which express forward activation and rebonding. In the above relations, δ, A, and G are experimental constants. The work term, w, can be described in resemblance to the kinetics concepts as

$$w_b = a_b \frac{\Delta K}{E} \quad \text{and} \quad w_{rb} = a_{rb} \frac{\Delta K}{E},$$

where α is an experimentally determined fracture mechanics parameter and E is the elastic modulus.

Figure 2.37 illustrates that equation (2.26) describes reference fatigue behavior for a wide variety of materials. The formal correspondence between equation (2.26) and the equations shown previously for thermally activated processes does not imply that reference fatigue is also thermally activated; it indicates, rather, that an exponential form represented by equation (2.26) and the corresponding expressions of $\mathscr{k}'_b$, $\mathscr{k}'_{rb}$ offer an attractive alternative to the power-function type fatigue relations.

Exponential relations of the type represented in equations (2.24) and (2.26) have been investigated for a very wide range of materials and fatigue regimes. In nearly all cases, they have proved valid. The fracture kinetics theory does offer a consistent descriptive system: the SCC process is defined fully by a theoretically rigorous expression; the effect of load variation is a rational extension of the SCC behavior, when rebonding is appropriately considered; and the inclusion of reference fatigue is a natural, although semi-empirical, extension of the rigorous formulation of the corrosion fatigue process [84—86].

Extensive investigations [57] of multiple fracture-mode processes that operate in parallel or in sequential kinetics have been carried out to evaluate temperature and stress intensity dependence. These studies aimed at predicting activation energies for corrosion fatigue and creep fatigue processes. It was noted that cracks growing by more than one fracture kinetics mode are quite common in fatigue [58] and time-dependent embrittlement processes [59]: for instance, in Fe-based systems low-temperature fatigue is controlled by parallel processes.

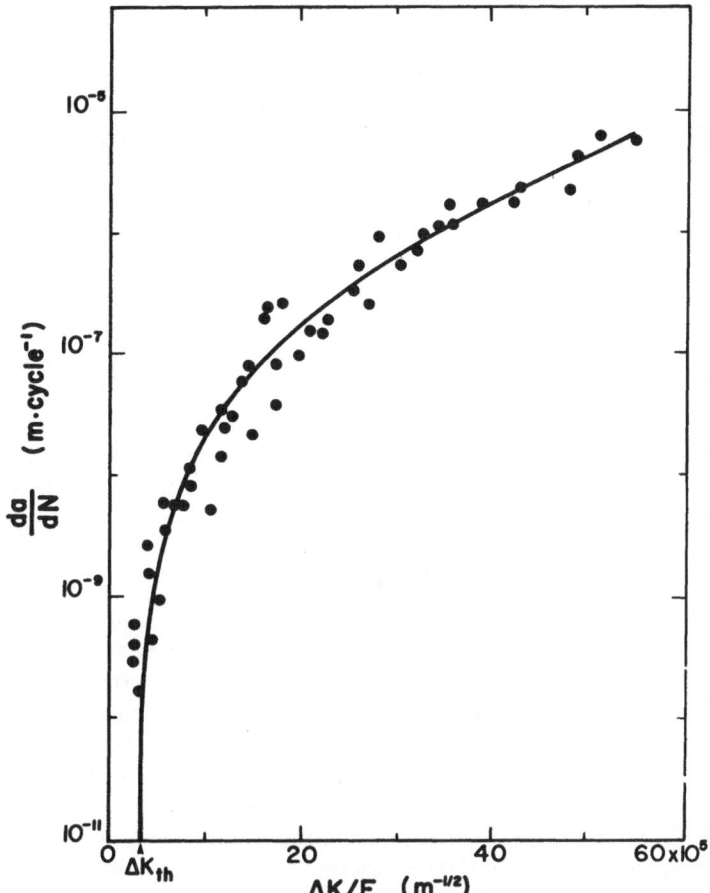

Figure 2.37. Reference fatigue crack growth rate versus modulus normalized stress intensity for Mg, Al, Ag, Cu, Zi, Ti, Fe, Ni, Be, Mo, alloys, some polymers. The solid line was calculated with equation (2.26) [46, 55]. Measurements were carried out at 21—24 °C, 1—5 Hz, in 10^{-5} Torr vacuum in sinusoidal loading, with $0.1 \leqslant R \leqslant 0$.

It is in this context that the environmental fatigue crack growth rate relation [45]

$$\left(\frac{\mathrm{d}a}{\mathrm{d}N} \right)_e = \left(\frac{\mathrm{d}a}{\mathrm{d}N} \right)_r + \left(\frac{\mathrm{d}a}{\mathrm{d}N} \right)_{cf} + \left(\frac{\mathrm{d}a}{\mathrm{d}N} \right)_{scc}$$

is assembled, with the understanding that each term on the right-hand side expresses the stress intensity by exponential relations that range from empirical to rigorously derived. One perceives a promise of unity here: the theoretical description of the full range of fatigue.

2.7. The revised Griffith theory

As noted in Chapter 1, the Griffith theory implies that the crack front moves as a rigid straight line. Under certain conditions this model provides a first approximation for the investigation of fracture behavior. Generally, however, it is too simple: it disregards the basic and universal physical fact that subcritical cracks never grow in such a manner because of (1) the presence of micro-structural imperfections, and (2) the character of thermal activation.

Microstructural imperfections

Microstructural imperfections are always present in materials. The simplest are vacancies — the absence of atoms from their geometrically defined lattice positions. It has been shown that vacancy concentration cannot be reduced below a minimum level determined by thermodynamic equilibrium; the very processes of plastic deformation always raise the concentration above this level.

Crystalline materials subjected to plastic deformation also contain line defects, dislocations. Although in thermodynamic equilibrium there could be no dislocations present, they are very difficult to remove.

Other common defects that influence crack growth include the surface imperfections formed by grain boundaries. Industrial materials always contain impurities as a natural consequence of standard production processes, and alloying elements intentionally added to the matrix material. These may be present in aggregates of varying sizes, and may be distributed regularly or randomly. But, in all cases, they cause variations in the atomic bond strength, an alternation of stronger with weaker resistance against crack growth.

Fabrication creates cavities and microcracks as well. These may range in size from the concentration of a few vacancies at the atomic, embryonic scale, to micropores, to cavities. All these imperfections — voids, microcrack regions, crazes and cavities — can be pre-existing, or they can develop from, and in turn influence, the crack growth process itself. Microcracks in structural components are now considered as inevitable; they must always be taken into account in the design and inspection scheduling of airplanes, pressure vessels and other struc-tures where safety standards are critical.

Directional variations of atomic strength are also present in most crystalline materials. These have, of course, a strong effect on the rate and direction of crack growth.

In all cases, microstructural effects are measured by the kinetics combination of elementary rate constants k, the activation energy ΔG^+, the work factor α, and the crack growth step length L. It is to be noted that the defect structure may obviously affect the stress—strain field on the larger scale, where con-tinuum mechanics models are still applicable, while the atomic displacement—force distribution affects crack growth on a finer scale where the essential bond-breaking process occurs. Consequently, the mechanical work, W, that is available along the crack front is more or less randomly distributed. Further-more, because the rate of bond breaking is defined by the term $\Delta G^+ - W$, the thermal energy needed for crack growth also varies along the front. These conditions may indeed create different rate-controlling mechanisms. For example, in Regions II and III of SCC, some segments of the crack front may move by a corrosion—bond-breaking mechanism and others by bond breaking into an uncorroded region because of different defect conditions [1, 8, 18, 27, 39, 60—62].

Thermal activation

Bond breaking occurs where the energy deficiency $\Delta G^+ - W$ is supplied by thermal activation. Recall from Chapter 1 that the atomic vibrations propagate through the solid with random amplitudes and directions. When a wave crest with sufficiently enlarged interatomic distances reaches the crack front and is superimposed on an atomic structure region already stretched by the applied force, the bonds break. The interatomic distance is related directly to the interatomic energy; the amplitude waves are also energy waves. The crack grows wherever the energy balance reaches the state

thermal energy peak $= \Delta G^+ - W$.

Thermal vibrations along a row of atoms which compose the crack front produce atomic displacements with irregular waveforms. As depicted in Figure 2.38, the thermal energy distribution may be visualized as a highly uneven surface in the crack plane. Since the thermal energy is available in a random pattern, the crack front cannot expand in a straight line.

In the discussion of microstructural imperfections it was shown that the defects cause random distribution of both ΔG^+ and W in the crack-tip zone. When the equally irregular thermal energy is superimposed, a random pattern of energy sum results. In the regions where $\Delta G^+ - W =$ thermal energy, the

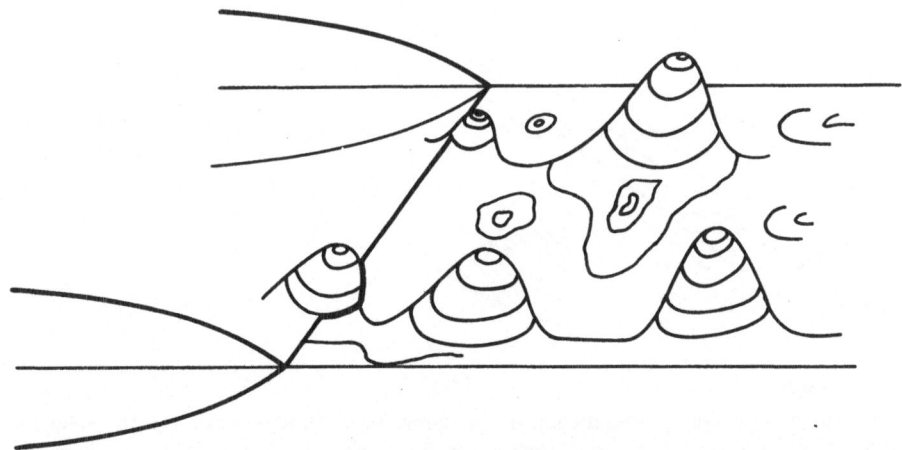

Figure 2.38. A schematic representation of the random thermal energy in the plane of the crack.

bonds are broken and the crack moves ahead. A statistical, probabilistic description is needed to represent this complex process.

The crack moves by one of two parallel processes. First, as shown in Figure 2.39, double kinks can form along the existing front at low bond-energy sites until they overlap and so the crack front moves into a new position. Alternatively, a few double kinks form and then spread sideways until they merge, as shown in Figure 2.40 [2, 8, 30, 65].

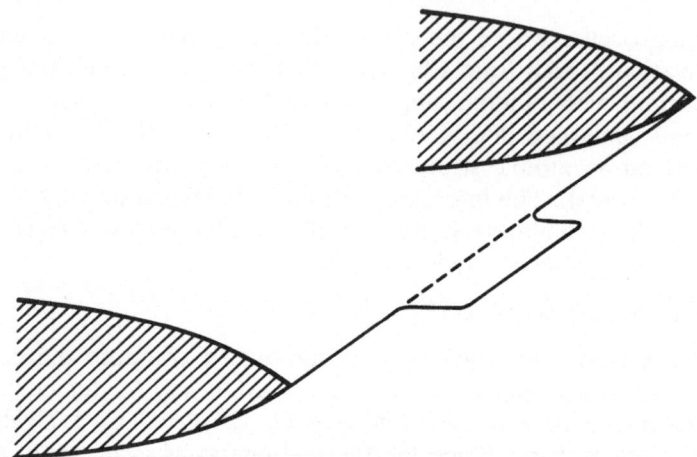

Figure 2.39. A schematic representation of the double-kink formation step. The broken line indicates the crack front before double-kink formation, while the solid line represents the crack front.

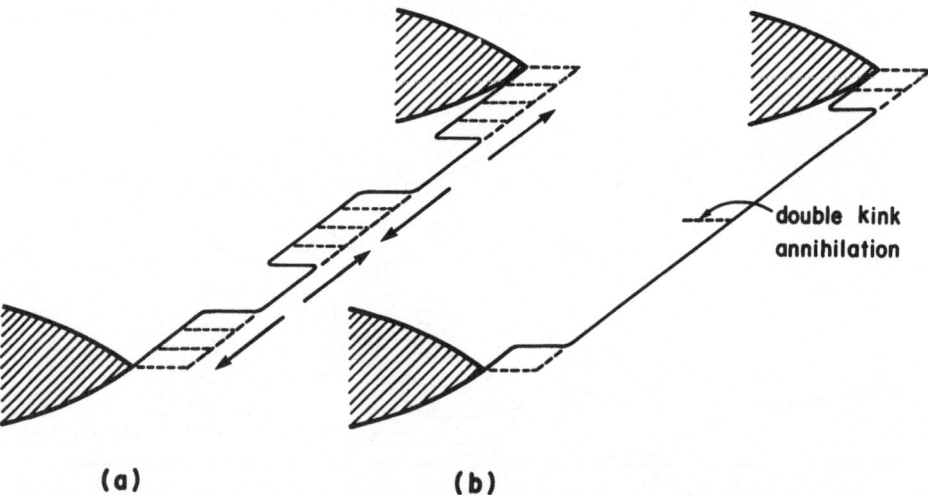

(a)   **(b)**

Figure 2.40. A schematic representation of the spreading of double kinks. (a) The solid lines indicate the configuration of the crack front during the spreading process. The broken lines show subsequent spreading steps, in the directions indicated by the arrows, until neighboring kinks merge. At that instant, a double kink is annihilated, as demonstrated in figure (b).

The formation of a kink — the initial step in the movement of the crack front — usually involves the simultaneous breaking of several bonds, a process requiring high energy. By contrast, sideways spreading steps may occur by bonds breaking one at a time, a relatively low-energy process. The probability of high-energy vibrational wave, and thus the probability of the arrival of such a configuration to the weakened site, is low: the probability of a double-kink formation step is consequently less than that of a spreading step, and it can be expected that this controls the crack growth rate.

There are, then, two mechanisms of crack growth:

(i) double-kink formation steps;

(ii) double-kink formation followed by spreading steps, a consecutive process.

The two may, of course, occur in parallel. The particular mechanism under investigation is defined by the specific microstructural factors, and loading and temperature service conditions. These are clarified rigorously and quantitatively by the constitutive equation [30, 64, 65].

The kinetics of the double-kink formation mechanism is straightforward. There is only a small probability that backward activation could heal an existing kink pair, because this would require the retracing of the breaking pattern — a rather unlikely event, rendered even less likely by the opposition of the mechanical work to healing. The kinetics, then, is simply

$$\mathcal{k} = \mathcal{k}_{dk} = \frac{kT}{h} \exp\left(-\frac{\Delta G_{dk}^+ - W_{dk}}{kT}\right),$$

where $\mathcal{k}$ is the average overall rate constant and $\mathcal{k}_{dk}$ is the double-kink formation rate.

The kinetics of the double-kink formation and consecutive spreading process can be expressed using the corresponding energy-barrier system (Figure 2.41). This consists of a double-kink formation barrier with breaking activation only, followed by a series of identical spreading steps with breaking and healing activation over each barrier.

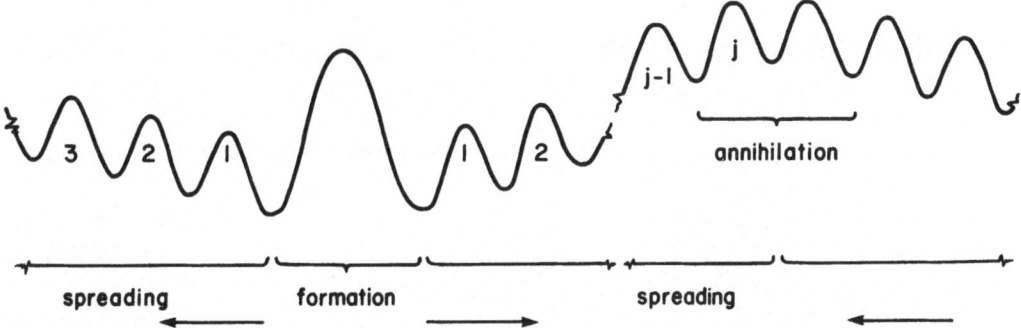

Figure 2.41. Schematic representation of a double-kink formation, spreading, annihilation energy-barrier system. Notice that the kinks spread in the two directions away from their formation barrier, as indicated by the arrows.

Because the spreading barriers are associated with one or a few atomic bonds only, they are low and consequently healing can take place. Even if the applied load is high, it still may not be sufficient to create the energy condition that would lead to $\Delta G_{sh}^{+}(W) \gg \Delta G_{sb}^{+}(W)$, with the corresponding distribution

$$\exp\left[-\frac{\Delta G_{sh}^{+}(W)}{kT}\right] \ll \exp\left[-\frac{\Delta G_{sb}^{+}(W)}{kT}\right],$$

where the subscripts 'sh' and 'sb' signify healing and breaking quantities in the spreading process, respectively (Figure 2.42).

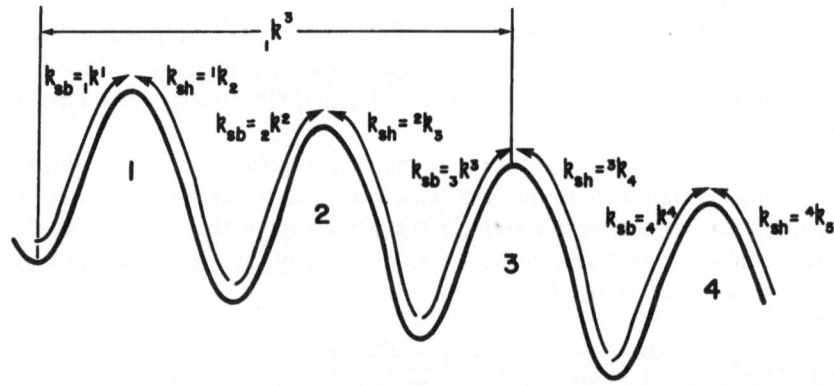

Figure 2.42. The symbol system used in the explanation of equation (2.27). The subscripts refer to the valley in front of the barrier designated by the number; the superscripts indicate the crest of the energy barrier identified by the number.

The constitutive equation that follows directly from the general expression of the kinetics equation is [30, 66]

$$v = L k = (L)\frac{1}{\dfrac{1}{k_{dk}} + \dfrac{1}{k_{dk}}\sum\limits_{j=1}^{j=n}\dfrac{1}{{}_{1}k^{j}} + \sum\limits_{i=1}^{i=n}\sum\limits_{j=0}^{j=n-i}\dfrac{1}{{}_{i}k^{i+j}}},$$

where the rate constant describes an activation combination originating from the lowest point of the barrier identified by the subscript and terminating at the crest, or activated state of the barrier, identified by the superscript.

Figure 2.42 helps to illustrate that the full expression of a rate constant, say ${}_{1}k^{3}$, is

$${}_{1}k^{3} = \frac{{}_{1}k_{2}\,{}_{2}k_{3}}{{}_{1}k^{1}\,{}_{2}k^{2}\,{}_{3}k^{3}}, \tag{2.27}$$

where, as usual,

$$k_{sb} = \frac{kT}{h}\exp\left(-\frac{\Delta G_{sb}^{+} - W_{sb}}{kT}\right)$$

and

$$\ell_{sh} = \frac{kT}{h} \exp\left(-\frac{\Delta G_{sh}^+ + W_{sh}}{kT}\right).$$

Because the energy barriers associated with the spreading steps can be considered as identical, it is sometimes useful, as a first approximation, to develop further the details of the constitutive equation. For this the literature can be consulted [30, 64—66]. These references show that under appropriate conditions, when the mechanism is controlled by double-kink formation, the equation can be reduced to the familiar expression

$$v = L \frac{kT}{h} \exp\left(-\frac{\Delta G^+ - W}{kT}\right).$$

Consequently, the typically observed crack velocity behavior expressed in this equation should not be interpreted as a simultaneous movement along the entire crack front; crack growth usually progresses by the disruption of individual segments and their subsequent merger [30].

A substantial theory has been developed for this important system due to its formal analogy with the Peierls—Nabarro double-kink nucleation and spreading kinetics of dislocation motion in plastic flow. Analyses are also available to determine the average distance between two double kinks, and other characteristics of the mechanism [67—72].

It is valid to assume here that the mechanism of crack growth processes may consist of the purely double-kink formation mechanism in parallel with the kink formation—spreading steps process, as noted before. The corresponding equation can be derived on the basis of developments described and referenced above; this is left as an exercise for readers experienced in kinetics analysis.

Attention is drawn to the lattice trapping theory that considers the discrete atomic nature of materials. Accordingly, the theory treats crack growth as a sequential overcoming of energy barriers [2, 8, 72—74].

2.8. Internal stress field and microstructural effects

The physical theory of fracture reveals that the process of crack growth consists of a sequence of atomic bond-breaking events, and therefore a function of the atomic-bond strength and the work contributed in the crack-tip region. It is well recognized that the internal stress field, on which the work depends, and the composition and structure that determine the bond strength of materials, significantly affect crack growth.

As distinguished from the residual stress, the internal stress field results from defects in crystalline materials — dislocations, vacancies, interstitials, impurity atoms, atoms of the alloying elements, segregations and particles embedded in the matrix. These imperfections distort the crystal lattice and thus change the interatomic distance which, in turn, alters the interatomic force: a stress field surrounds the crystal defect.

This internal stress field, σ_i, is superposed on the applied stress, σ_a, and the crack moves under their combined action: $\sigma_r = \sigma_a + \sigma_i$. The crack also encounters variations in bond strength as a result of inclusions, segregations, second phases, and textures such as eutectics. As the crack moves from one grain to another, differences in crystallographic orientation cause it to change direction, creating a varying applied stress field in the path of the crack.

The internal stress field, and the variations in the bond strength and in the crystallographic orientation, present a distribution of both crack driving force and resistance that is sometimes regular, but more often random. For quantitative analyses these effects must be incorporated in the mathematical expression of the crack velocity. For clarity of presentation the simple constitutive equation

$$v = L\ell = L\,\frac{kT}{h}\,\exp\left(-\frac{\Delta G^* - W}{kT}\right)$$

is considered [75].

As mentioned before, during growth the crack tip encounters variations in the atomic bond strength: ΔG^+ is therefore the function of the crack-tip location, formally described as

$$\Delta G^+ = \Delta G^+(x) \tag{2.28}$$

where x is the distance in the direction of the crack movement. The crack tip is actually under the effect of the resultant stress σ_r, a function of the internal stress field variation, expressed formally as

$$\sigma_r = \sigma_a(x) + \sigma_i(x). \tag{2.29}$$

The work, W, available for bond breaking depends on the stresses at the tip and on the orientation of the crack; as the crack moves through the grains, the stress intensity factor K also changes. The work is expressed as

$$W(\sigma_r, x) = \alpha K(\sigma_a, x) + \beta\sigma_i(x), \tag{2.30}$$

and the stress intensity factor K as

$$K(\sigma_a, x) = Y(x)\sigma_a(x)a^{1/2}, \tag{2.31}$$

where, as usual, Y is the function of the loading and geometrical boundary condition and, therefore, of the crack orientation, and $2a$ is the crack size. Substituting equations (2.28) to (2.31), the crack velocity is expressed for thermally activated, time-dependent brittle fracture as

$$v = L\,\frac{kT}{h}\,\exp\left[-\frac{\Delta G^*(x) - \alpha Y(x)\sigma_a(x)a^{1/2} - \beta\sigma_i(x)}{kT}\right]. \tag{2.32}$$

Each of the three terms, $\Delta G^+(x)$, $Y(x)$, and $\sigma_i(x)$ can be randomly distributed, or have essentially periodic distribution functions.

Consider first the variation of one of the three terms, the internal stress. A regular sinusoidal distribution of σ_i, leading to a simple form of equation (2.32) is assumed; other forms, including random variation, can also be incorporated following the method discussed in this section. The crack velocity, $\bar{v}$, measured at the macroscopic scale, is the average of the instantaneous velocity, v, over the cycle length, Z, of the internal stress field

$$\bar{v} = \frac{Z}{t},$$

where t is the period of crack growth of a cycle. From the definition of the instantaneous velocity

$$v = \frac{dx}{dt},$$

the period, t, is

$$\int_{t(x)}^{t(x+Z)} dt = \int_{x}^{x+Z} \frac{dx}{v},$$

and hence

$$\bar{v} = \frac{Z}{\int_{x}^{x+Z} \frac{dx}{v}}, \qquad (2.33)$$

where

$$v = L \frac{kT}{h} \exp\left(-\frac{\Delta G^{+} - \alpha Y\sigma_{a}a^{1/2} - \beta A \sin(2\pi/Z)x}{kT}\right). \qquad (2.34)$$

The harmonic-function representation of the internal stress field is useful: any other physically reasonable stress distribution can be expressed by an appropriate combination of harmonic functions using Fourier analysis. A similar consideration applies to the analysis of the effect of the variation of $\Delta G^{+}(x)$ and $K(x)$.

A full description of the average crack velocity is obtained here in a functional form of the integral in equation (2.33). Alternatively, however, numerical solutions must be developed: often, simplified stress-field representations can be valuable models, as the following example shows.

EXAMPLE 2.7. To illustrate how kinetics analysis can be applied to determine the effects of the internal stress field on crack velocity, a specimen geometry is considered for which $Y\sigma_{a}a^{1/2} = $ constant. With this, equation (2.34) becomes

$$v = L\ell^0 \exp\left(\beta A \frac{\sin(2\pi/Z)x}{kT}\right). \tag{2.35}$$

The sinusoidal expression can be approximated with convenient models, such as the triangular or rectangular waveforms shown in Figure 2.43.

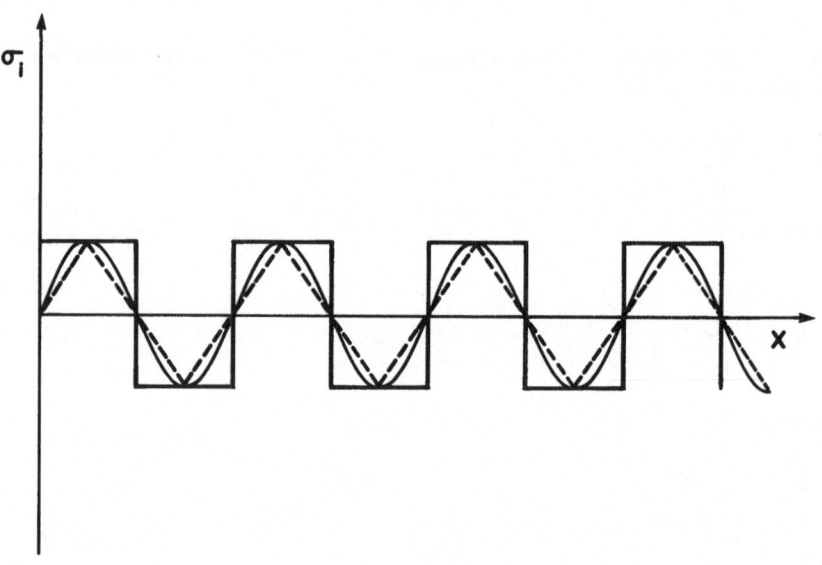

Figure 2.43. The triangular and rectangular waveforms for the approximation of the sinusoidal variation of internal stress.

Using the simple rectangular approximation in equation (2.35), the average crack velocity is [75]

$$\bar{v} = \frac{Z}{\dfrac{1}{L\ell^0}\left[\displaystyle\int_0^{Z/2}\exp\left(-\frac{\beta A}{kT}\right)dx + \int_{Z/2}^{Z}\exp\left(\frac{\beta A}{kT}\right)dx\right]}$$

$$= \frac{LZ\ell^0}{\dfrac{Z}{2}\exp\left(-\dfrac{\beta A}{kT}\right) + \dfrac{Z}{2}\exp\left(\dfrac{\beta A}{kT}\right)}$$

$$= \frac{L\ell^0}{\cosh\dfrac{\beta A}{kT}}. \tag{2.36}$$

Equation (2.36) offers several interesting conclusions. First, it shows that the larger the amplitude of the internal stress field, the slower the average macroscopic crack velocity, even though stress regions above and below the applied stress are involved symmetrically. That is, while the internal stress has equal effects in hindering and driving the crack, the net effect of these two processes is to slow down crack growth, rather than to balance it out.

Secondly, because $\mathscr{k}^0$ increases and cosh $\beta A/kT$ decreases with increasing temperature, the equation indicates that the higher the temperature, the faster the crack grows. It also follows that at high internal stresses the sensitivity of the macroscopic crack velocity to the temperature increases. As well, analysis of the model shows that the wavelength of the internal stress field has no effect on the average crack velocity.

In equation (2.32) the three terms $\Delta G^+(x)$, $Y(x)$, and $\sigma_i(x)$, representing variations in structure, crack direction and internal stress field, are of equal character; thus, equation (2.36) is equally valid for all three effects. Consequently, the conclusions derived from the effects of the internal stress field apply to the effects of structure and crack direction as well. In fact, for some of the structural effects the rectangular model is physically preferable to the triangular or sinusoidal formulations. Two examples in particular are worth noting:

(1) The atomic-bond strength varies periodically in eutectic solids between high and low values of ΔG^+. For lamellar eutectics this variation can be represented by a rectangular dependence of ΔG^+ on x.

(2) The direction of microscopic crack growth, which varies from grain to grain but remains constant within the boundaries of the grain, can be expressed through the fracture mechanics factor Y. This value changes abruptly when the crack crosses the grain boundary, but remains constant between. Again, a rectangular model appropriately represents the physical process.

2.9. Lifetime analysis

Investigations of crack growth behavior are directed ultimately at finding dependable, scientific ways to determine and improve the lifetimes of components subject to failure by fracture [37].

This is a central problem in the development of engineering design, operating procedures, and maintenance tests and schedules. The analysis is involved, even for structural and machine components subjected to constant load and temperature and a stable environment. The interaction effects of fluctuating loads render the determination of crack growth behavior very complex and not yet fully tractable; variations in temperature and environment pose additional problems. Fracture kinetics, with its rigorous description of the effects of load, temperature, microstructure, and environment, can provide guidance in the challenging task of lifetime analysis [76].

Generally, a crack starts at an initial size a_0 and grows at a subcritical velocity v until the stress factor reaches the critical value. The crack then propagates at a very high velocity. Because the critical velocity is of the order of several hundred meters per second, this last period of subcritical crack growth is usually very short; it is the initial period that essentially determines the lifetime. In the following discussion, the total crack growth period will be analyzed using fracture kinetics.

By definition, the velocity is defined as $v = da/dt$, and the growth of the crack from size a_1 to a_2 is

$$\int_{a_1}^{a_2} \frac{1}{v} \, da = \int_{t_1}^{t_2} dt,$$

that is, the growth time between two crack sizes is

$$t_2 - t_1 = \int_{a_1}^{a_2} \frac{1}{v(a)} \, da \tag{2.37}$$

showing that the crack velocity must be described in terms of the crack size.

Fracture kinetics can be applied to obtain a rigorous derivation of velocity. First, the effect of the microstructure on the mechanism is determined and the values of the material characteristics, ΔG^+, α, and also L, are measured, as discussed before. Once the constitutive equation is established, the velocity expression can be integrated. In many cases, this procedure is rather straightforward; for others, conceptual or mathematical manipulations may be needed to develop a functional form. The procedure always provides powerful information on the effects of applied load, component geometry, temperature, environment and microstructure.

The following example uses the simplest kinetics to explain the general principles of lifetime determination. It also illustrates a conceptual manipulation that is often essential to the analysis and requires, more often than not, a measure of ingenuity. If these efforts fail, one can fall back on the numerical solution of the lifetime integral represented in equation (2.37).

EXAMPLE 2.8. Consider the simplest kinetics description for forward activation when $W = \alpha K$:

$$v = L \frac{kT}{h} \exp\left(-\frac{\Delta G^+ - \alpha K}{kT}\right),$$

where

$$K = Y\sigma a^{1/2},$$

and hence,

$$\int_{t_0}^{t_c} dt = \int_{a_0}^{a_c} \frac{1}{L} \frac{h}{kT} \exp\left(\frac{\Delta G^+ - \alpha Y\sigma a^{1/2}}{kT}\right) da. \tag{2.38}$$

However, this form cannot be readily integrated: the $a^{1/2}$ factor creates difficulties. As discussed before, under many conditions the K^2 dependence of crack velocity is as useful as the K dependence; in some cases it is critically better. This concept can now be introduced in equation (2.38), resulting in

$$\int_{t_0}^{t_c} dt = \frac{1}{L} \frac{h}{kT} \exp\left(\frac{\Delta G^+}{kT}\right) \int_{a_0}^{a_c} \exp\left(-\frac{\alpha Y^2 \sigma^2 a}{kT}\right) da.$$

When the initial time $t_0 = 0$, and Y is independent of the crack size, the lifetime is

$$t_c = -\frac{h}{L\alpha Y^2 \sigma^2} \exp\left(\frac{\Delta G^+ - \alpha Y^2 \sigma^2 a}{kT}\right)\Bigg|_{a_0}^{a_c}.$$

Under most conditions the effect of the pre-exponential factor is negligible compared to that of the exponent and can be considered as a constant. The lifetime is then

$$t_c = A\left[\exp\left(\frac{\Delta G^+ - \alpha Y^2 \sigma^2 a_0}{kT}\right) - \exp\left(\frac{\Delta G^+ - \alpha Y^2 \sigma^2 a_c}{kT}\right)\right]. \tag{2.39}$$

Equation (2.39) represents explicitly the effects of stress, component and crack geometry, initial crack size, temperature and microstructure; it is a valuable tool for the design of quantitative tests to determine material characteristics. It is also widely applicable. For instance, each of the three regions of stress corrosion cracking is described individually by this type of expression. The total lifetime is simply the sum of the crack growth times in each region. Furthermore, because near the critical crack size $\Delta G^+ \simeq \alpha Y^2 \sigma^2 a_c$; the approximation

$$t_c \simeq A \exp\left(\frac{\Delta G^+ - \alpha Y^2 \sigma^2 a_0}{kT}\right)$$

is valid.

This expression lends itself readily to interpretations. To note but one example, the well-known concept of temperature-compensated lifetime can be expressed as

$$kT \ln\left(\frac{t_c}{A}\right) = \Delta G^+ - \alpha Y^2 \sigma^2 a_0 \tag{2.40}$$

where the temperature effect is now isolated and the lifetimes at different temperatures are brought into common, normalized dependence represented on the right-hand side of equation (2.40). This provides an important design relation when plotted as the function of the initial crack size or the stress. Quite conveniently, the initial crack size dependence is a straight line, and the stress dependence — a second-order parabola — is also very simple.

Lifetime analyses for mechanisms with more complex kinetics can be carried out using this example.

A well-known series of tests and evaluations is described in detail below. Because full experimental results are given, the series also serves as an evaluation database and provides practice in problem-solving. A full set of answers is provided.

EXAMPLE 2.9. A wide range of metallic, ceramic and polymeric materials were investigated in a coherent test series [77, 78]. A first set of analyses was carried out with the bond-breaking rate constant only, and the results were represented in a lifetime versus stress coordinate system. In this study the mechanical energy was considered to vary linearly with the normal stress, σ and the crack size a. Hence the crack growth rate is expressed as

$$\frac{da}{dt} = L \, \frac{kT}{h} \, \exp\left(-\frac{\Delta G_b^+ - \alpha_b \sigma a}{kT}\right)$$

and the lifetime is

$$\int_{a_0}^{a_c} \exp\left(-\frac{\alpha_b \sigma a}{kT}\right) da = L \, \frac{kT}{h} \, \exp\left(-\frac{\Delta G_b^+}{kT}\right)(t_c - t_0),$$

where a_0 and a_c are the initial and critical crack sizes; the work was considered to be $\alpha \sigma a$; and t_0 and t_c are the initial time and the time at fracture.

When $t_0 = 0$,

$$\frac{kT}{\alpha_b \sigma} \left[\exp\left(-\frac{\alpha_b \sigma a_0}{kT}\right) - \exp\left(-\frac{\alpha_b \sigma a_c}{kT}\right) \right] = t_c L \, \frac{kT}{h} \, \exp\left(-\frac{\Delta G_b^+}{kT}\right)$$

and

$$t_c = \frac{h}{L \alpha_b \sigma} \left[\exp\left(\frac{\Delta G_b^+ - \alpha_b \sigma a_0}{kT}\right) - \exp\left(\frac{\Delta G_b^+ - \alpha_b \sigma a_c}{kT}\right) \right]. \qquad (2.41)$$

The stress dependence of failure time was measured for 50 materials that included metals, ceramics and polymers, in constant stress loading. A comparison between the experimental results and the calculated behavior at various temperatures is shown in Figure 2.44 for AgCl, Al and Plexiglas.

The apparent activation energy under the discussed conditions was determined using a semi-empirical method. Figure 2.44 illustrates that for each of the materials tested, the lifetime lines for all temperatures meet at the same stress. The critical time corresponding to this point was considered to be equal to the pre-exponential factor $h/L\alpha_b \sigma$; knowing this, the apparent activation energy, $\Delta G^+(W)$ was defined from equation (2.41). It was found that for all of the tested crystalline and polymeric materials, the stress dependence of $\Delta G^+(W)$ was linear, as predicted by equation (2.41). Figure 2.45 and 2.46 illustrate

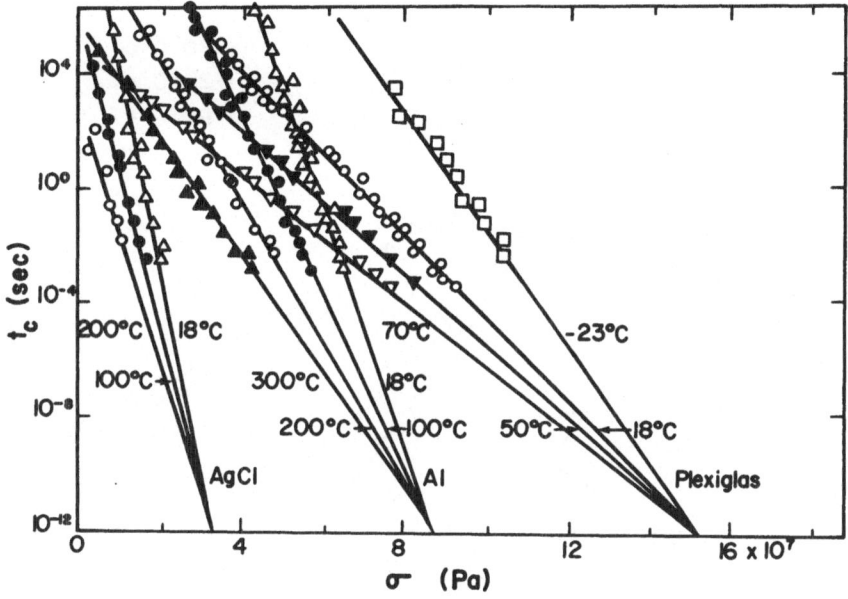

Figure 2.44. The stress and temperature dependence of the lifetime of AgCl, Al, and Plexiglas [77].

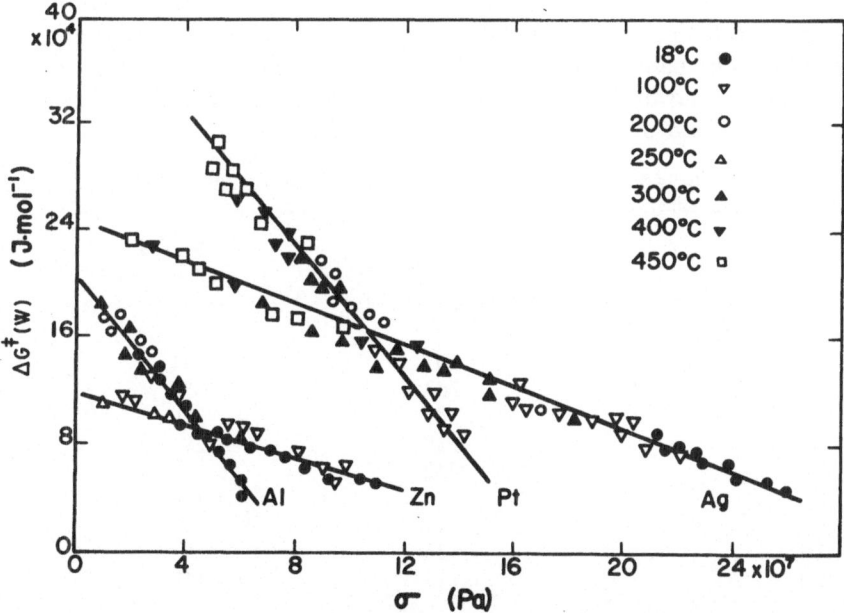

Figure 2.45. The stress dependence of the apparent activation energy at different temperatures for polycrystalline Al, Zn, Pt and Ag [77].

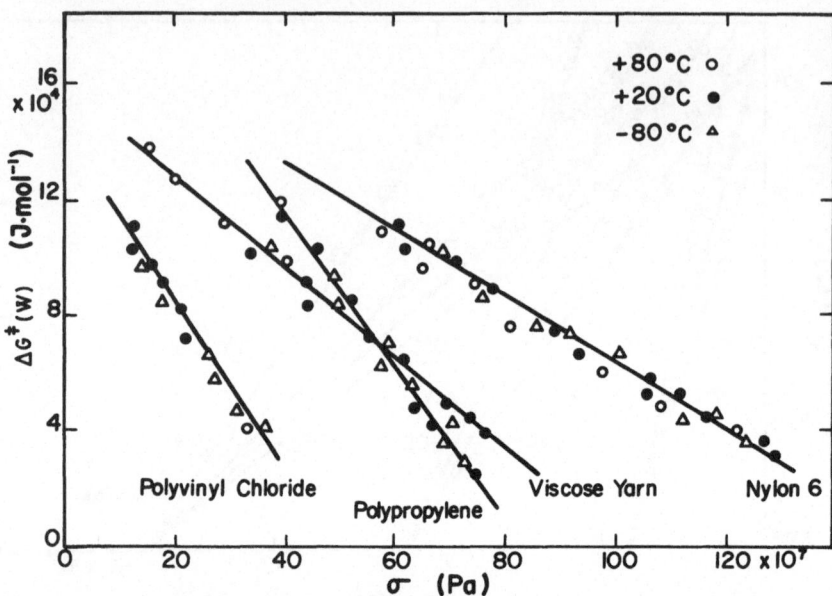

Figure 2.46. The stress dependence of the apparent activation energy at different temperatures for polyvinyl chloride, viscose yarn, polypropylene and Nylon 6 [77].

this behavior in the high-stress region. The figures show that $\Delta G^+(W)$ was independent of temperature, in agreement with theory. Because the stress dependence of the apparent activation energy is linear, the activation energy ΔG^+ was evaluated by extrapolation to $\sigma = 0$. The activation energy values obtained in this manner are shown in Table 2.3 for polycrystalline metals and single crystals, and in Table 2.4 for polymers, together with the corresponding atomic-bond energies ΔG^+_{bond}.

Table 2.3. The measured activation energies and the bond energies for metals.

Metals	ΔG^+ $\times 10^3\,\mathrm{J\,mol^{-1}}$	ΔG^+_{bond} $\times 10^3\,\mathrm{J\,mol^{-1}}$
Polycrystalline		
Pt	501	531
Ni	364	355
Cu	343	339
Ag	259	284
Al	222	—
Pb	176	199
Mg	142	152
Zn	125	—
Cd	117	112
Single crystal		
Al	225	230
Zn	146	132

Table 2.4. The measured activation energies and the bond energies for polymers.

Polymers	ΔG^+ $\times 10^3$ J mol^{-1}	ΔG^+_{bond} $\times 10^3$ J mol^{-1}
Teflon	313	318–334
Polypropylene	234	230–242
Polystyrene	226	230
Polymethyl methacrylate	226	217–227
Nylon	188	180
Polyvinyl chloride	146	134

At the maximum rate of bond breaking, when $\alpha_b \sigma a_c = \Delta G^+$, the mechanical work is large enough so that thermal energy is not needed and the fracture process is not thermally activated. At low stresses the measurements showed that the fracture life is longer than predicted by equation (2.41), indicating that at these stresses the bond-healing process was not negligible (Figure 2.47).

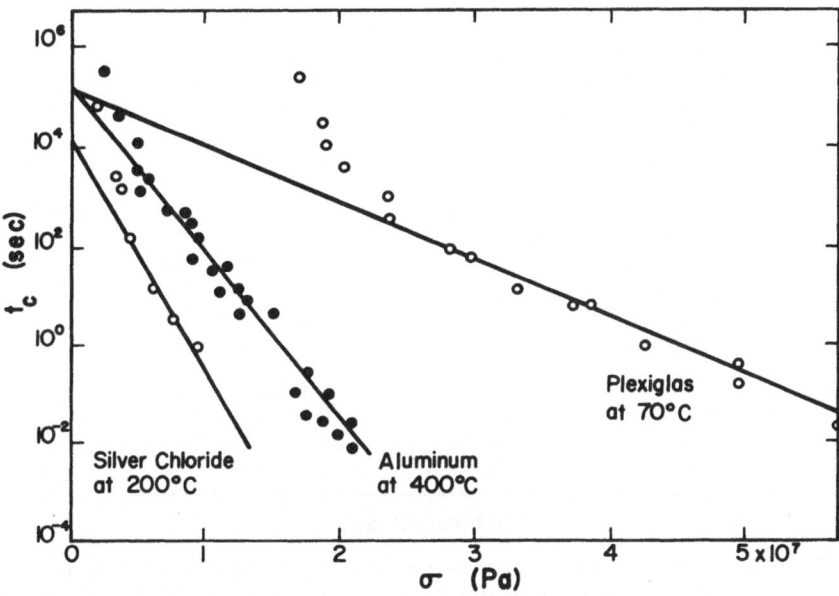

Figure 2.47. The figure illustrates the difference between the lifetime calculated from equation (2.41) and the fracture time measured for AgCl, Al, and Plexiglas. The lines were calculated and the symbols indicate the measured values [77].

The constitutive equation (equation (2.41)) was extended by considering that the rate of bond breaking may be different from the rate of bond healing. The crack velocity is then

$$\frac{da}{dt} = (L)(k_b - k_h). \tag{2.42}$$

Considering again that the force acting on a bond is related to the stress as expressed before, it was found that the logarithm of fracture time is related to the stress, as shown in Figure 2.48. The figure illustrates that the fracture theory in the form of equation (2.42) represents well the behavior observed over both stress ranges. The derivation of the constitutive equation is left as an exercise; for comparison, reference [78] can be consulted.

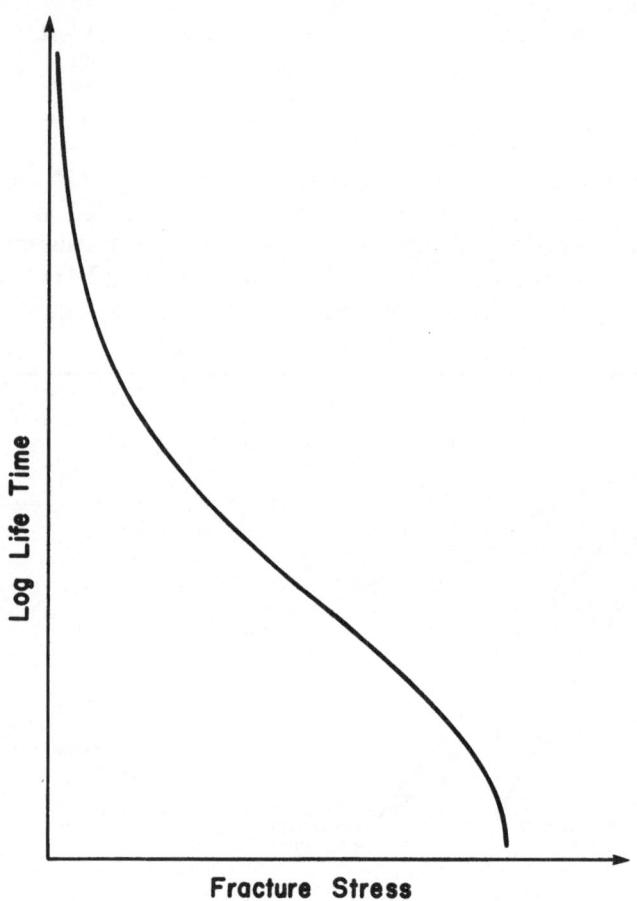

Fracture Stress

Figure 2.48. The typical shape of the stress dependence of lifetime when bond mending is considered, as expressed by equation (2.42).

2.10. Comments and summary

The selected applications of fracture kinetics illustrated the theory developed in Part 1, providing guidance in solving engineering and research problems. Above all, Part 2 offers an opportunity to practice the development of fracture constitutive equations — from doing comes knowledge.

Environment-assisted cracking, which includes stress corrosion cracking as a special case, is one important area of subcritical, thermally activated fracture. The EAC process involves the simultaneous action of microstructural degradation and mechanical load, resulting in crack growth that would not occur in a non-reactive environment. Its two major components are: (a) structural deterioration of the material; and (b) crack growth into the deteriorated, weakened region.

(a) Structural deterioration, caused by (external) degradation, involves several processes, which may consist of these steps: transport of the agent to the crack-tip zone; physical adherence of the agent to the surface; establishment of the chemical reactions; separation of the reaction products from the surface; transport of the by-products from the reaction zone.

(b) Bond breaking also involves various processes. It was established that crack growth occurs by formation of a double kink along the crack front, and subsequent sideways spreading activation. Because energy conditions favor this type of crack front progression, it has a higher probability of occurring than a long crack-front growth mechanism.

The full description of the EAC mechanism is complex, and, when it is required, the mathematical processing is difficult. However, the full descriptions may often be replaced by their simplified forms. To do so, the model should consider only those elements that are essential contributors to the particular conditions. A number of mechanisms have been recognized and their fracture kinetics descriptions developed [14]: the reaction kinetics of SCC and of corrosion fatigue were reviewed recently [17, 18, 79].

For illustrative purposes, this chapter employed a model representing SCC as controlled, in Region I, by stress-enhanced chemical bond breaking, and in Region II by a largely stress-independent degrading transport step, such as diffusion. Both processes were rather arbitrarily simplified to 'forward and backward' steps of breaking and healing; and the degrading step and its reversal, in terms of the corresponding rate constants k_b, k_h, k_d, k_r. The model was not meant as a generally valid statement of the kinetics of environment-assisted fracture. Indeed, radiation degradation does not include the classical chemical processes of corrosion. For expediency of presentation, the examples were simplified to consider only an elementary form of chemical reaction, diffusion, and bond-breaking activation, as the controlling processes. In fact, for each actual application, the corresponding mechanism, and thus the model, must be specifically established — generally a most difficult task that requires the combined expertise of solid mechanics, materials science and chemistry [45].

The concentration of the corrosive component in the environment is of obvious importance. The consideration of chemical effects and diffusion processes is outside the scope of this presentation; these, and some other effects are not discussed. For the corresponding extension of the kinetics analysis of environment-assisted crack growth the references provide guidance [6, 8, 14, 16, 33]. It is to be noted that chemical reactions and diffusion are thermally activated processes of which crack growth is 'one of their kind'; their inclusion follows the (same) concepts elaborated upon for fracture kinetics. The great degree of complexities involved in these processes require specialized attention. The description often takes the form

$$\text{rate} \propto f(\text{composition, concentration}) \frac{kT}{h} \exp\left[-\frac{\Delta G^*(W)}{kT}\right]$$

[8, 33, 43, 44, 55]. The rate constant should be considered with these extensions. The transport of reactive environment experiencing viscous drag in fluid flow has been incorporated in recent works [8, 33].

Problems associated with high-temperature fracture and corrosion fatigue are often investigated and described with power-function relations. These may provide analytically advantageous constitutive laws, but their validity as rigorous, rational expressions is not necessarily established for many engineering applications. Furthermore, it is important to realize that because these are empirical or semi-empirical formulations, their use for extrapolation is dangerous. A physical interpretation of the parameters of these equations was introduced.

This chapter also revised, from the fracture kinetics point of view, the modifications and extensions of the classical Griffith theory summarized in Chapter 1. This revision included consideration of the double-kink formation and sideways spreading mechanism of crack growth. The influence of entropy on fracture kinetics analyses were recognized [16, 80, 83]; entropy effects in their probabilistic form are considered in Chapter 3.

Polymeric materials are extensively used, even in structural applications: for these, time-dependent fracture has an important role [1, 81, 82]. Their analysis by thermal activation theory is essential [39].

The constitutive laws must express, rigorously and quantitatively, the effects of the microstructure. This, and the effects of the internal stress field on subcritical crack propagation, were discussed.

The fracture kinetics theory of constitutive laws leads to rational determination of the lifetime of components. The analysis is well defined from the theory: complex mechanical and thermal loads, as well as microstructural conditions, can be described quantitatively. The physical processes themselves may lead to mathematical difficulties that prevent closed form or analytical solutions. Computer-assisted numerical methods provide the lifetime in such cases [81, 87].

Part 2 has considered a variety of applications in terms of deterministic constitutive laws. It has demonstrated constitutive laws that are derived from the physical process of subcritical crack growth, composed of bond-breaking steps controlled by random thermal vibration. The process is essentially probabilistic: the deterministic applications must be considered as the average behavior.

Probabilistic fracture kinetics theory and constitutive laws

PART 1
PROBABILISTIC FRACTURE KINETICS THEORY

It was shown in Chapters 1 and 2 that the physical process of crack growth is the net effect of the random breaking and healing of atomic bonds; that the randomness itself results from the stochastic fluctuations of the atomic vibrational amplitude; and that this fluctuation is, in fact, the thermal energy that controls the rate of atomic breaking and healing steps. It was also demonstrated that the rate theory of statistical mechanics rigorously describes the average rate of the random breaking and healing activations by the elementary rate constant k. Because the rate of activations associated with any macroscopic crack movement is very large — $10^8\,\mathrm{s}^{-1}$, give or take a few orders of magnitude — random fluctuation is not perceived on this scale. Nevertheless, it has very significant and often essential consequences on crack growth.

As will be seen, the deterministic crack velocity often equals the expectation value of the real, probabilistic behavior. However, because the probabilistic nature of the processes controlling fracture were inadequately understood, statistical variations were usually assigned to experimental conditions, and to the non-reproducible character of the microstructure of the material, and to its heterogeneity; hence the statistical consideration of the $S-N$ curves of fatigue, and the Weibull statistics. Both these techniques are widely used and are essential tools for designers and operating engineers. Indeed, structural irregularity is an ever-present reality, but it should not serve as a blanket explanation for all random effects [1—3]*.

It has been established since the mid-1970s that because bond-breaking and healing events are always random, time- and temperature-dependent subcritical crack growth is always probabilistic: the statistical variation caused by material microstructure is superposed on the fundamentally random physical process. In some cases, the effects of microstructure overshadow this basic physical effect, while in others the physical effect is more significant.

* References to Chapter 3, Part 1, are given on page 143.

Chapter 2, Part 1, dealt with the deterministic character of the kinetics theory, while Part 2 focused on the applications of the deterministic constitutive laws of crack growth. This chapter now draws a full distinction between deterministic and probabilistic fracture kinetics, and provides an introduction to the latter model; the discussion is supplemented by the presentation of some important applications of the probabilistic description.

The mathematical apparatus used to develop the concepts in this chapter may not be entirely familiar to some readers. Whenever possible, methods are employed that do not require a specialized mathematical background. However, at certain points, the reader will be asked to follow some rather elaborate mathematical processes. This effort will be found to yield considerable practical benefits in the design of lifetime and maintenance of machinery and structural components.

Although the authors attempted to render this chapter accessible to those encountering the subject for the first time, some concepts may present difficulties. We suggest that the chapter be read through first, with careful attention to the Examples. After that, a re-reading of the section 'Consecutive systems' will be rewarding.

3.1. The theory of probabilistic fracture kinetics

As discussed previously, crack growth results from a random sequence of bond-breaking and healing events. This is the physical essence of every thermally activated process, all of which are addressed by an important branch of physics: the theory of random walk, or Brownian motion. The theory provides important insights into the essential characters of probabilistic crack growth.

It is customary to introduce the random walk concept by drawing an analogy with the progress of a drunken sailor — or, in this case, let us say a fracture specialist — who is heading from Tavern A to Tavern B. This undertaking occurs in discrete steps, each randomly spaced in time and at utterly unpredictable frequencies of forward and backward staggerings and sequences. But, under the effect of a (more or less helpful) drive, the frequency (or probability) of forward steps towards Tavern B can be greater than the frequency of backward staggerings. Predicting the probability of the fracture specialist arriving at Tavern B at a set distance from Tavern A then becomes an algebraic counting process.

Similarly, the crack staggers from the initial size to the critical size by a random sequence of forward (bond-breaking) steps, and backward (bond-healing) steps. In this random process some cracks make more forward steps than others over the same time: they grow to a larger size. Consequently, under the effect of the same stress intensity, and over the same time, cracks grow to different sizes. The randomness of the physical process always results in a crack-size distribution: the crack growth behavior is always probabilistic.

A rigorous mathematical development of the random walk theory of crack growth is given in Part 2. It is shown there that under simple conditions the crack size distribution is binomial (Figure 3.1). The deterministic crack size, at time t, is the expectation value; cracks of smaller and larger sizes do develop with probabilities according to the binomial distribution function [4].

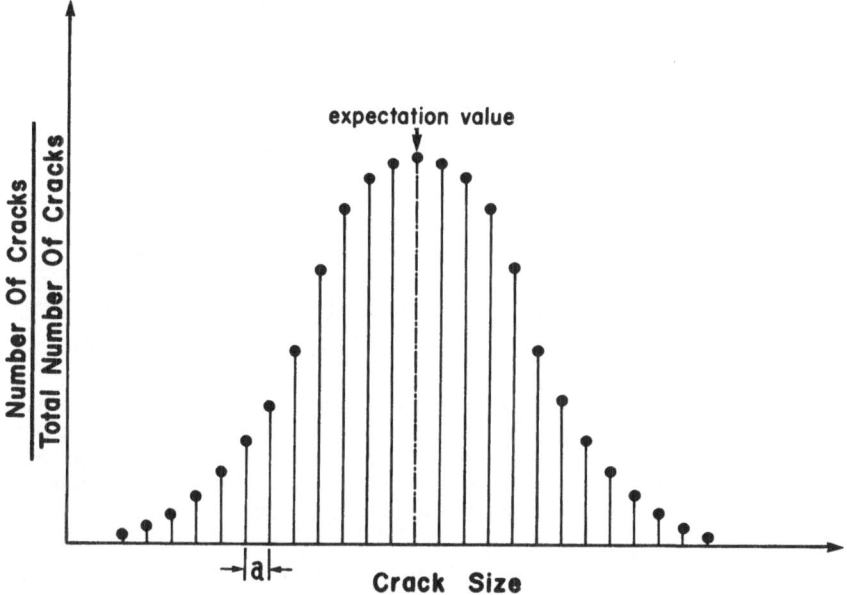

Figure 3.1.   Crack-size distribution as the consequence of the random frequency of bond-breaking and healing steps; at time t and with constant stress factor.

Random walk mathematics is usually not a practical tool for crack growth analyses. In its place the Markov-chain probability method is used; an alternative, the Fokker—Planck differential equation of transport processes, is also discussed in Part 2.

The division of fracture kinetics into three basic types, discussed in association with the deterministic theory — consecutive systems, parallel systems and consecutive systems combined in parallel — are valid for probabilistic kinetics as well. In fact, the probabilistic interpretation is highly consistent, and thus conceptually satisfying: it will be shown in Part 2 that it includes the deterministic formulation as a special case.

Probabilistic fracture kinetics is best approached by considering first its application to consecutive systems.

Consecutive systems

All crack growth processes consist of a sequence of bond-breaking and occasional bond-healing steps that involve consecutive rearrangements of the atomic configuration. This sequence is shown in Figures 3.2(a) and 3.3(a); Figures 3.2(b) and 3.3(b) represent schematically the concepts used in the mathematical

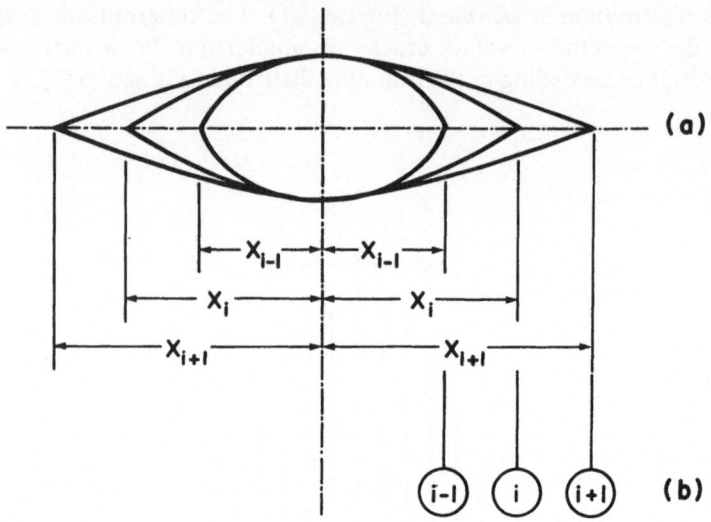

Figure 3.2.   Schematics for the illustration of probabilistic crack tip movements. (a) illustrates the condition of cracks of size x_i; smaller cracks, at x_{i-1}, grow to this size by a single activation; cracks of size x_{i+1} shrink to this size by a single activation; and cracks of x_i grow larger or smaller than this size at the rates of k_b and k_h, respectively. (b) illustrates schematically the stations of the Markov-chain representation that correspond to sizes x_{i-1}, x_i and x_{i+1}.

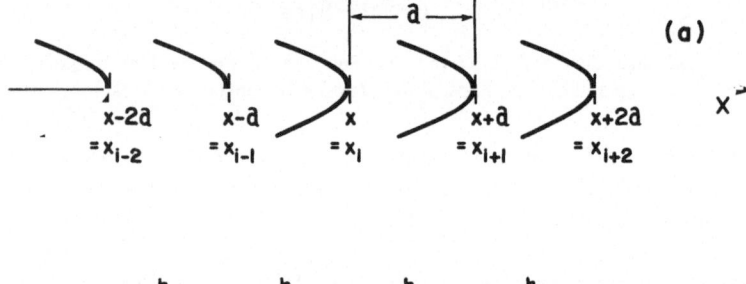

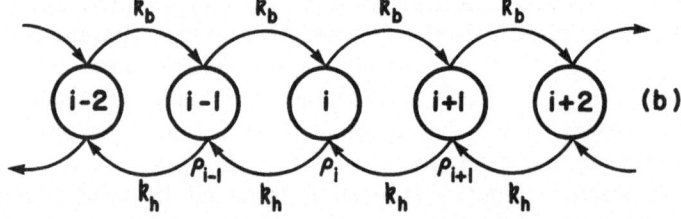

Figure 3.3.   (a) The successive positions of the crack tip; (b) the schematic for the representation of the crack-tip positions and the rate of bond breaking and healing. Note that crack sizes can differ only by an integer multiple of the interatomic distance a.

formulation. Cracks of size $x = x_i$ in Figure 3.3(a) are represented by position i in Figure 3.3(b). The cracks in Figure 3.3(a) grow to size $x + a = x_{i+1}$ by bond breaking at k_b frequency, and shrink to the size $x - a = x_{i-1}$ by healing at k_h frequency. These processes are identified in Figure 3.3(b) by the positions i, $i +$

1, and $i - 1$, respectively. The arrows represent the frequencies of the transitions between positions as $\mathcal{k}_b$ and $\mathcal{k}_h$. Figure 3.3(b) is, in fact, a standard illustration of the Markov-chain method of probability theory [4, 5].

As previously noted, crack movements occur in the sequence illustrated in Figures 3.3(a) and (b). The Markov theory thus provides a powerful method for the analysis of the crack growth if the physical conditions agree with the conditions of the Markov theory:

(1) the process must be random;
(2) the rate constants (or, in the terminology of the Markov theory, the transition coefficients) must be independent of time;
(3) the rate constants must depend on the crack size only, and not on the particular breaking and healing sequences that brought the crack to that size.

Condition (2) is not a severely limiting factor: mathematical apparatus exists that accepts time-dependent transmission coefficients but their discussion is outside the scope of this book.

It should be noted also that condition (3) does not imply that structural changes and other history-dependent processes cannot be considered in the analysis. They can be, and are expressed by the appropriate value of the rate constants. It means, rather, that each rate constant is a single-valued function of the crack size: a definite value of the rate constant is associated with a specific crack size. This is, of course, in agreement with the implied fracture mechanics definition of the stress factors $K, J, \mathcal{G}, C^*$, etc. [6, 7].

To develop the Markov-chain description of crack growth, it is useful to visualize an experiment that tests a very large number of specimens with cracks of various sizes. At the instant t, the number of cracks of size $x = x_i$ is ρ_i; of size $x - a = x_{i-1}$ is ρ_{i-1}; of size $x + a = x_{i+1}$, ρ_{i+1}; and so on. The corresponding states in the Markov chain are represented schematically in Figure 3.3(b) as ρ_i cracks in state i; ρ_{i-1} cracks in state $i - 1$; etc. Consider now the change in the number of cracks ρ_i in state i during the time interval t to $t + \Delta t$:

$$\rho_i(t + \Delta t) - \rho_i(t).$$

This change occurs because cracks both grow and shrink. Some of the ρ_{i-1} cracks will grow from size x_{i-1} to size x_i, while some of the ρ_{i+1} cracks will shrink from x_{i+1} to x_i. At the same time, some of the ρ_i cracks of size x_i will grow from x_i to x_{i+1} and others will shrink to x_{i-1}. The process can be visualized, therefore, as a flow of crack sizes into and out of state i. Table 3.1 summarizes these movements, the associated crack sizes, and the corresponding states in the Markovian model shown in Figure 3.3(b). Accordingly, the change in the number of cracks of size $x = x_i$ during the t to $t + \Delta t$ time period is (in the following, L is omitted for ease of reading)

$$\rho_i(t + \Delta t) - \rho_i(t) = \mathcal{k}_b \rho_{i-1} \Delta t - (\mathcal{k}_b + \mathcal{k}_h)\rho_i \Delta t + \mathcal{k}_h \rho_{i+1} \Delta t. \qquad (3.1)$$

Hence the Markov theory states that the rate of change in the number of

Table 3.1. Summary of the number of cracks changing during the time interval t to $t + \Delta t$.

Initial crack size	Corresponding state	Number of cracks moving	Final crack size	Corresponding state	Direction of change
x_{i-1}	$x_i - a$	$k_b \rho_{i-1} \Delta t$	x_i	x_i	$\rightarrow$
x_i	x_i	$k_b \rho_i \Delta t$	x_{i+1}	$x_i + a$	$\rightarrow$
x_i	x_i	$k_h \rho_i \Delta t$	x_{i-1}	$x_i - a$	$\leftarrow$
x_{i+1}	$x_i + a$	$k_h \rho_{i+1} \Delta t$	x_i	x_i	$\leftarrow$

cracks of x_i size, or at state i, is

$$\frac{\rho_i(t + \Delta t) - \rho_i(t)}{\Delta t} = k_b \rho_{i-1} - (k_b + k_h)\rho_i + k_h \rho_{i+1}. \tag{3.2}$$

In the limit, as $\Delta t \to 0$, equation (3.2) becomes

$$\frac{d\rho_i}{dt} = k_b \rho_{i-1} - (k_b + k_h)\rho_i + k_h \rho_{i+1}.$$

A similar expression can be written for all but the initial and critical crack sizes. The complete Markov-chain model that includes these sizes, or states, is shown in Figure 3.4. Here, healing from the initial size x_0 is impossible, because the

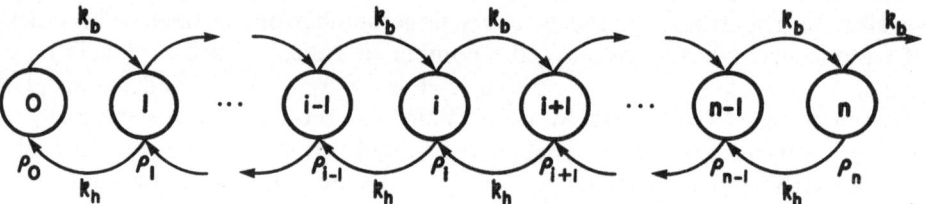

Figure 3.4. The Markov-chain model of crack propagation from the initial size (state 0) to final size (state n), through the intervening sizes of type i. Again, each state represents a crack size greater by one interatomic distance, a, than the previous one. The number of cracks at each size, or state, is $\rho_0, \rho_1, \ldots, \rho_{i-1}, \rho_i, \rho_{i+1}, \ldots, \rho_{n-1}, \rho_n$, respectively.

crack cannot become smaller than x_0; furthermore, since x_0 is the smallest, the initial size, there can be no growth into x_0. Healing cannot occur into the critical size x_n because the next state of the Markov chain is final failure, the parting of the specimen. Accordingly, the first equation is

$$\frac{d\rho_0}{dt} = -k_b \rho_0 + k_h \rho_1$$

while the last is

$$\frac{d\rho_n}{dt} = -k_b \rho_n + k_b \rho_{n-1} - k_h \rho_n.$$

The equation system that fully describes the behavior can then be assembled:

$$\frac{d\rho_0}{dt} = -k_b\rho_0 + k_h\rho_1$$

$$\frac{d\rho_1}{dt} = k_b\rho_0 - (k_b + k_h)\rho_1 + k_h\rho_2$$

$$\vdots$$

$$\frac{d\rho_i}{dt} = k_b\rho_{i-1} - (k_b + k_h)\rho_i + k_h\rho_{i+1} \tag{3.3}$$

$$\vdots$$

$$\frac{d\rho_{n-1}}{dt} = k_b\rho_{n-2} - (k_b + k_h)\rho_{n-1} + k_h\rho_n$$

$$\frac{d\rho_n}{dt} = k_b\rho_{n-1} - (k_b + k_h)\rho_n.$$

For clarity of presentation, the stress intensity factor was considered to be constant during crack growth from the initial to the critical size. Usually, however, the stress factors depend on the crack size and the microstructure may change during crack growth. Thus, more often than not, the rate constants reflect the crack size. Figure 3.5 illustrates the Markov chain in its general form. Each rate constant depends on the size from which the crack grows or shrinks; the figure identifies the breaking- and healing-rate constants accordingly. The general form of the differential equation system is

$$\frac{d\rho_0}{dt} = -k_{b0}\rho_0 + k_{h1}\rho_1$$

$$\frac{d\rho_1}{dt} = k_{b0}\rho_0 - (k_{b1} + k_{h1})\rho_1 + k_{h2}\rho_2$$

$$\vdots$$

$$\frac{d\rho_i}{dt} = k_{bi-1}\rho_{i-1} - (k_{bi} + k_{hi})\rho_i + k_{hi+1}\rho_{i+1} \tag{3.4}$$

$$\vdots$$

$$\frac{d\rho_{n-1}}{dt} = k_{bn-2}\rho_{n-2} - (k_{bn-1} + k_{hn-1})\rho_{n-1} + k_{hn}\rho_n$$

$$\frac{d\rho_n}{dt} = k_{bn-1}\rho_{n-1} - (k_{bn} + k_{hn})\rho_n.$$

It is usually assumed that the solution of a linear differential equation system

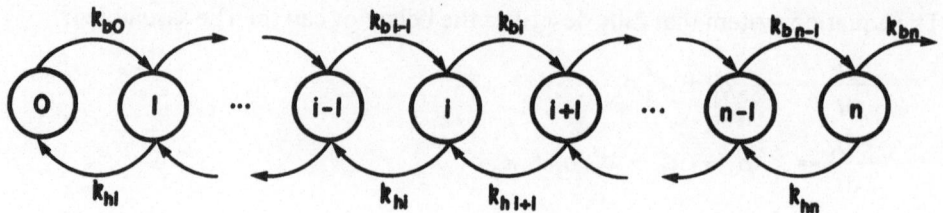

Figure 3.5.   The generalized schematic representation of the Markov-chain model.

can be written as

$$\rho_i = \sum_{j=1}^{j=n+1} C_{ji} \exp(-\lambda_j t) + C_i. \tag{3.5}$$

Substitution of equation (3.5) into the differential equation system (3.4) produces a system of algebraic equations of the type

$$\frac{d}{dt} \sum_{j=1}^{j=n+1} C_{ji} \exp(-\lambda_j t) = k_{bi-1} \left[\sum_{j=1}^{j=n+1} C_{ji-1} \exp(-\lambda_j t) + C_{i-1} \right] -$$

$$- (k_{bi} + k_{hi}) \left[\sum_{j=1}^{j=n+1} C_{ji} \exp(-\lambda_j t) + C_i \right] +$$

$$+ k_{hi+1} \left[\sum_{j=1}^{j=n+1} C_{ji+1} \exp(-\lambda_j t) + C_{i+1} \right]. \tag{3.6}$$

In equation (3.6) the time-independent terms must satisfy the relation

$$k_{bi-1} C_{i-1} - (k_{bi} + k_{hi}) C_i + k_{hi+1} C_{i+1} = 0.$$

For the time-dependent terms each $e^{-\lambda_j t}$ factor must satisfy equation (3.6): hence

$$\frac{d C_{ji} \exp(-\lambda_j t)}{dt} = k_{bi-1} C_{ji-1} \exp(-\lambda_j t) -$$
$$- (k_{bi} + k_{hi}) C_{ji} \exp(-\lambda_j t) +$$
$$+ k_{hi+1} C_{ji+1} \exp(-\lambda_j t),$$

and thus the general term is

$$k_{bi-1} C_{ji-1} - (-\lambda_j + k_{bi} + k_{hi}) C_{ji} + k_{hi+1} C_{ji+1} = 0.$$

The λ_j eigenvalues are derived from the condition that the determinant of the coefficients must be zero

$$
\begin{vmatrix}
-k_{b0} & k_{h1} & 0 & 0 & \cdots & & \\
k_{b0} & \lambda - k_{b1} - k_{h1} & k_{h2} & 0 & \cdots & & \\
\vdots & \vdots & \vdots & \vdots & & & \\
& & & & \vdots & & \\
\cdots & 0 & k_{bn-2} & \lambda - k_{bn-1} - k_{hn-1} & k_{hn} \\
\cdots & 0 & 0 & k_{bn-1} & \lambda - k_{bn} - k_{hn}
\end{vmatrix} = 0.
$$

The determinant results in the characteristic equation $D = 0$ with $n + 1$ roots: $\lambda_1, \ldots, \lambda_n, \lambda_{n+1}$.

When all roots are distinct, the crack size distribution is expressed by equation (3.5). When the expansion of the determinant results in multiple roots, then the crack size distribution is expressed by a polynomial form. For example, if $\lambda_1 = \lambda_2 = \lambda_3$ then the number of crack tips at the ith valley is

$$
\rho_i = (C_{1i} + C_{2i}t + C_{3i}t^2)\exp(-\lambda_1 t) +
$$

$$
+ \sum_{j=4}^{j=n+1} C_{ji}\exp(-\lambda_j t) + C_i.
$$

Substitution of the $n + 1$ roots into equation (3.6) produces a system of algebraic equations where all λ and k are defined: since at the initial time $t = 0$

$$
\rho_i(t = 0) = \sum_{j=1}^{j=n+1} C_{ji} + C_i,
$$

knowing the initial distribution the remaining C_{ji} parameters can be determined. Thus a system of parameters is obtained for each eigenvalue λ_j.

This analysis provides all the information needed to understand and describe consecutive crack growth processes within the context of probabilistic fracture kinetics. The following simple examples will aid the understanding of the mathematical procedures.

EXAMPLE 3.1. Part 2 will show that, under many practical conditions, the Markov chain takes the reduced form shown in Figure 3.6.

The corresponding differential equation system is

$$
\frac{d\rho_0}{dt} = -k_{b0}\rho_0 + k_{h1}\rho_1, \tag{3.7a}
$$

$$
\frac{d\rho_1}{dt} = k_{b0}\rho_0 - (k_{b1} + k_{h1})\rho_1 \tag{3.7b}
$$

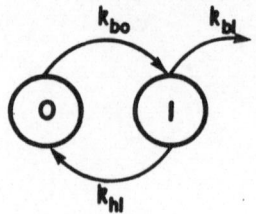

Figure 3.6.   Markov-chain representation of a simple crack growth process.

and the solution is

$$\rho_0 = C_0 + C_{10}\exp(-\lambda_1 t) + C_{20}\exp(-\lambda_2 t), \tag{3.8a}$$

$$\rho_1 = C_1 + C_{11}\exp(-\lambda_1 t) + C_{21}\exp(-\lambda_2 t). \tag{3.8b}$$

Substitution of equations (3.8) into equation (3.7a) results in

$$-C_{10}\lambda_1\exp(-\lambda_1 t) - C_{20}\lambda_2\exp(-\lambda_2 t)$$
$$= -k_{b0}C_0 - k_{b0}C_{10}\exp(-\lambda_1 t) - k_{b0}C_{20}\exp(-\lambda_2 t) +$$
$$+ k_{h1}C_1 + k_{h1}C_{11}\exp(-\lambda_1 t) + k_{h1}C_{21}\exp(-\lambda_2 t), \tag{3.9a}$$

and equations (3.8) into equation (3.7b) results in

$$-C_{11}\lambda_1\exp(-\lambda_1 t) - C_{21}\lambda_2\exp(-\lambda_2 t)$$
$$= k_{b0}C_0 + k_{b0}C_{10}\exp(-\lambda_1 t) + k_{b0}C_{20}\exp(-\lambda_2 t) -$$
$$- (k_{b1} + k_{h1})C_1 - (k_{b1} + k_{h1})C_{11}\exp(-\lambda_1 t) -$$
$$- (k_{b1} + k_{h1})C_{21}\exp(-\lambda_2 t). \tag{3.9b}$$

Equations (3.9a) and (3.9b) split into time-independent and time-dependent terms.

The time-independent equations are

$$-k_{b0}C_0 + k_{h1}C_1 = 0 \quad \text{and} \quad k_{b0}C_0 - (k_{b1} + k_{h1})C_1 = 0;$$

hence $C_0 = 0$ and $C_1 = 0$.

The time-dependent terms must also satisfy equations (3.9a) and (3.9b) for λ_1 and λ_2 separately. For the λ_1 eigenvalue association there are two equations

$$(\lambda_1 - k_{b0})C_{10} + k_{h1}C_{11} = 0, \tag{3.9c}$$

and

$$k_{b0}C_{10} + (\lambda_1 - k_{b1} - k_{h1})C_{11} = 0. \tag{3.9d}$$

Similar relations are obtained for λ_2: these algebraic equations must satisfy the determinant, which being the same for λ_1 and λ_2, is written as

$$\begin{vmatrix} \lambda - k_{b0} & k_{h1} \\ k_{b0} & \lambda - k_{b1} - k_{h1} \end{vmatrix} = 0.$$

Hence

$$(\lambda - k_{b0})(\lambda - k_{b1} - k_{h1}) - k_{b0}\, k_{h1} = 0,$$

and so

$$\lambda^2 - \lambda(k_{b0} + k_{b1} + k_{h1}) + k_{b0}\, k_{b1} = 0.$$

The two roots, λ_1 and λ_2, can be readily established. Substitution of λ_1 into equations (3.9c) or (3.9d) produces an algebraic equation system in terms of the parameters C_{10} and C_{11}:

$$C_{11} = \frac{k_{b0} - \lambda_1}{k_{h1}}\, C_{10}. \qquad (3.10a)$$

A similar procedure yields

$$C_{21} = \frac{k_{b0} - \lambda_2}{k_{h1}}\, C_{20}. \qquad (3.10b)$$

Finally, the solution is obtained by substituting equations (3.10a), (3.10b), and $C_0 = 0$, $C_1 = 0$ into equations (3.8a) and (3.8b). In the resulting descriptions, C_{10} and C_{20} are defined by the initial conditions. At $t = 0$,

$$\rho_0(t = 0) = C_{10} + C_{20},$$

$$\rho_1(t = 0) = \frac{k_{b0} - \lambda_1}{k_{h1}}\, C_{10} + \frac{k_{b0} - \lambda_2}{k_{h1}}\, C_{20}. \qquad (3.11)$$

EXAMPLE 3.2. Figure 3.7 represents the three-state Markov chain, a somewhat more complicated system.

The differential equation system is

$$\frac{d\rho_0}{dt} = -k_{b0}\, \rho_0 + k_{h1}\, \rho_1,$$

$$\frac{d\rho_1}{dt} = k_{b0}\, \rho_0 - (k_{b1} + k_{h1})\rho_1 + k_{h2}\, \rho_2,$$

$$\frac{d\rho_2}{dt} = k_{b1}\, \rho_1 - (k_{b2} + k_{h2})\rho_2.$$

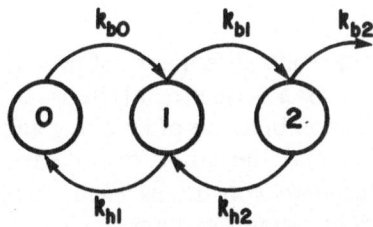

Figure 3.7.   A simple Markov-chain equivalent to a crack growth process from the initial size (state 0), to the final, critical size (state 2), when the material parts.

The solution is expressed as

$$\rho_0 = C_0 + \sum_{j=1}^{3} C_{j0} \exp(-\lambda_j t),$$

$$\rho_1 = C_1 + \sum_{j=1}^{3} C_{j1} \exp(-\lambda_j t),$$

$$\rho_2 = C_2 + \sum_{j=1}^{3} C_{j2} \exp(-\lambda_j t).$$

Substitution produces three systems, one for each eigenvalue λ_j. For example, λ_1 is expressed as

$$-C_{10}\,\lambda_1 \exp(-\lambda_1 t) = -k_{b0}\,C_0 + k_{h1}\,C_1 + [-k_{b0}\,C_{10} + k_{h1}\,C_{11}]\exp(-\lambda_1 t),$$

$$-C_{11}\,\lambda_1 \exp(-\lambda_1 t) = k_{b0}\,C_0 - (k_{b1} + k_{h1})\,C_1 + k_{h2}\,C_2 +$$
$$+ [k_{b0}\,C_{10} - (k_{b1} + k_{h1})\,C_{11} + k_{h2}\,C_{12}]\exp(-\lambda_1 t),$$

$$-C_{12}\,\lambda_1 \exp(-\lambda_1 t) = k_{b1}\,C_1 - (k_{b2} + k_{h2})\,C_2 +$$
$$+ [k_{b1}\,C_{11} - (k_{b2} + k_{h2})\,C_{12}]\exp(-\lambda_1 t).$$

Separation of the time-dependent and time-independent terms produces the system

$$(\lambda_1 - k_{b0})\,C_{10} + k_{h1}\,C_{11} = 0,$$

$$k_{b0}\,C_{10} + (\lambda_1 - k_{b1} - k_{h1})\,C_{11} + k_{h2}\,C_{12} = 0,$$

$$k_{b1}\,C_{11} + (\lambda_1 - k_{b2} - k_{h2})\,C_{12} = 0.$$

This can be completed as an exercise.

Clearly, the modest increase in the length of the Markov chain in Example 3.2 introduces considerable complexities. Computer programs are available that greatly reduce the time and effort required to arrive at numerical solutions.

Parallel systems

In accordance with physical reality, the theory of probabilistic kinetics considers all crack growth processes as a sequence of steps. In this context, a parallel system is actually a parallel combination of breaking and healing sequences with the simplifying assumption that the stress factor varies only from branch to branch. By definition, this then constitutes each of the parallel branches: in reality, this is what is meant by parallel systems.

Each branch in the parallel system is described by a set of differential equations of the type shown in equation (3.3), where the rate constants are different and operate independently in each of the branches.

Consecutive systems combined in parallel

In the probabilistic description of fracture kinetics, this becomes a special case of parallel systems. Here, the breaking and healing rates are not constant in each branch, but vary as described by the system of differential equation (3.4).

This description is valid for the majority of cases. Occasionally, however, other mechanisms operate, as in a system with interacting parallel branches, or with a crack concentration ρ_i^n with $n > 1$: these occurrences are rare and rather difficult to deal with; consequently, they are of interest mainly to specialists.

3.2. Comments and summary

The differential equation systems described above yield probabilistic constitutive equations for crack growth processes. Applications of particular interest are where it is required
  (1) to define the number of cracks of a specific size at any specified time during testing or the service life of a material, from $\rho_0 + \rho_1 + \cdots + \rho_{i-1} + \rho_i + \rho_{i+1} + \cdots + \rho_{n-1} + \rho_n = \rho_t$, where ρ_t is the total number of cracks;
  (2) to specify the crack velocity of any of the crack sizes at the instant t;
  (3) to determine the average crack velocity during growth from the initial to critical size.
Probabilistic fracture kinetics is the appropriate representation of the process of random thermal activation that controls subcritical crack growth. Deterministic kinetics provides the expectation value of the probabilistic description. Often, when the variance (the scatter about the average velocity) is small, this is a good approximation. The definition of deterministic kinetics implied that the crack velocity reached a steady-state condition; the probabilistic theory represents this state and the transition state as well [6—13] as is shown in Part 2 of this chapter.

Because the probabilistic description represents the random character of the physical process of crack growth, it has far-reaching validity. However, mathematically it is more complex: a differential equation system is associated with the rate-controlling mechanism. It can be solved in a closed form, or by numerical methods using Markov-chain mathematics. Obviously, practical considerations often dictate the choice between the deterministic and probabilistic theories. However, the relatively complex manipulations of the probabilistic model are rewarded by a more powerful descriptive and predictive tool. Part 2 of this chapter describes certain cases in which application of this theory may indeed be essential to efficient design and maintenance engineering.

PART 2
PROBABILISTIC CONSTITUTIVE LAWS

Probabilistic fracture kinetics considers that crack growth is a thermally activated, time- and temperature-dependent process controlled by the random thermal fluctuation of atoms. In Part 1, using the Markov-chain analysis, a differential equation system was derived that is represented by a typical component as

$$\frac{d\rho_i}{dt} = k_{bi-1}\rho_{i-1} - (k_{bi} + k_{hi})\rho_i + k_{hi+1}\rho_{i+1},$$

where ρ_i is the number of cracks of size a_i; t is the time; and the subscripts $i-1$, i, and $i+1$ indicate that the quantity belongs to the a_{i-1}, a_i and a_{i+1} crack sizes, respectively. Because the frequency of the steps is randomly distributed the crack sizes are also probabilistic.

For Part 2, the examples have been chosen to illustrate the essential concepts and the mathematical apparatus as simply as possible. The discussion progresses gradually from elementary to more complex applications. It is advisable to proceed systematically, paying equally close attention to the mathematical apparatus used in processing and to the conceptual aspects of the applications. The reader will be rewarded by mastery of a method that is, to be sure, far from elementary, but which provides insight into the process of subcritical crack growth and a powerful, widely applicable tool for design and maintance.

3.3. Applications of probabilistic kinetics

Breaking activation only: $k_h = 0$. In the simplest case of crack growth, all breaking rate constants, k_b, are equal and healing is negligible. The material properties are thus homogeneous and unaltered by crack growth, and the stress intensity factor, K, is constant. This condition is of rather limited practical interest; it can only be approximated within a narrow range, under laboratory conditions. However, it provides an excellent introduction to the understanding of probabilistic behavior.

The Markov chain is represented in Figure 3.8. The corresponding differen-

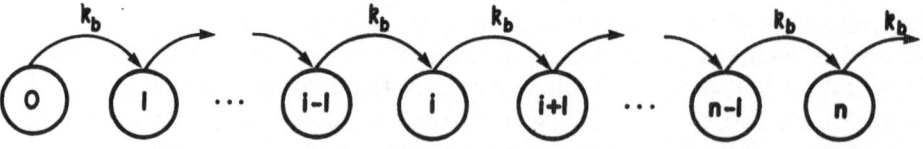

Figure 3.8. The Markov-chain representation of crack propagation with breaking activation only and k_b = constant.

tial equation system is

$$\frac{d\rho_0}{dt} = -k_b \rho_0$$

$$\frac{d\rho_1}{dt} = k_b \rho_0 - k_b \rho_1$$

$$\frac{d\rho_2}{dt} = k_b \rho_1 - k_b \rho_2$$

$$\vdots \qquad\qquad (3.12)$$

$$\frac{d\rho_i}{dt} = k_b \rho_{i-1} - k_b \rho_i$$

$$\vdots$$

$$\frac{d\rho_{n-1}}{dt} = k_b \rho_{n-2} - k_b \rho_{n-1}$$

$$\frac{d\rho_n}{dt} = k_b \rho_{n-1} - k_b \rho_n.$$

While the general solution developed in the previous chapter is applicable, other mathematical techniques are often more useful. Consider that the first equation in (3.12) can be directly integrated [1—3]*

$$\rho_0 = \rho_t \exp(-k_b t) \qquad\qquad (3.13)$$

where the constant ρ_t, being the total number of cracks, signifies that initially all cracks are in the ρ_0 position. After substitution of equation (3.13), the second equation of (3.12) is

$$\frac{d\rho_1}{dt} = k_b \rho_t \exp(-k_b t) - k_b \rho_1,$$

which, in turn, can be integrated by parts, resulting in

$$\rho_1 = \rho_t k_b t \exp(-k_b t).$$

Substitution into the third equation results in

$$\frac{d\rho_2}{dt} = \rho_t k_b^2 t \exp(-k_b t) - k_b \rho_2,$$

* References to Chapter 3, Part 2, are given on page 143.

which then yields

$$\rho_2 = \rho_t \frac{(k_b t)^2}{2!} \exp(-k_b t).$$

Proceeding in the same manner produces the well-known Poisson type relation

$$\rho_i = \rho_t \frac{(k_b t)^i}{i!} \exp(-k_b t).$$

Note that the calculated number of cracks ρ_i of size a_i is valid if the total number of cracks ρ_t is large and constant.

An alternative expression is often useful: the probability P_i of observing a crack of size a_i. Hence, for the case under discussion,

$$P_i = \frac{(k_b t)^i}{i!} \exp(-k_b t) = \frac{\rho_i}{\rho_t}. \tag{3.14}$$

These results illustrate an important advantage of probabilistic kinetics. The present example, and subsequent applications, will show that the descriptions provided by this physically realistic model lead to a wealth of information available from the probabilistic mathematics of transport process physics.

The Poisson distribution approaches the binomial distribution when the number of events, that is, the average number of crack growth steps, $k_b t$ is large. Indeed, this approximation is already a very close one after only a few steps: Figure 3.9 illustrates that when $k_b t > 6$ the distribution already takes the typical binomial form [3].

The following highly simplified example helps to illustrate the behavior after a small number of crack growth steps. Consider that each activation advances the crack by one atomic distance, say a $= 2 \times 10^{-8}$ cm, and that the apparent activation energy is $\Delta G^+(W) = 0.5$ eV. By substituting these quantities into the constitutive equation, the crack velocity at room temperature is found to be

$$v = a \frac{kT}{h} \exp\left[-\frac{\Delta G^+(W)}{kT}\right]$$

$$\cong 12 \times 10^{-8} \times 10^{12} \exp(-20) \text{ cm s}^{-1}$$

$$= 2.4 \times 10^{-4} \text{ cm s}^{-1}.$$

This is a typical value for Region I stress corrosion behavior in metals and ceramics. The corresponding rate constant is $k_b = 1.2 \times 10^4 \text{ s}^{-1}$, and the $(k_b t) = 6$ state is reached in about 5×10^{-4} s — a very short time indeed.

Because the distribution can be approximated by the normal distribution, the probability of the occurrence of a particular crack size is [1—3]

$$P = \left(\frac{1}{2\pi a^2 k_b t}\right)^{1/2} \exp\left[-\frac{(x - a k_b t)^2}{2a^2 k_b t}\right].$$

Accordingly, the average crack size (the expectation value) is $2a k_b t$ and the

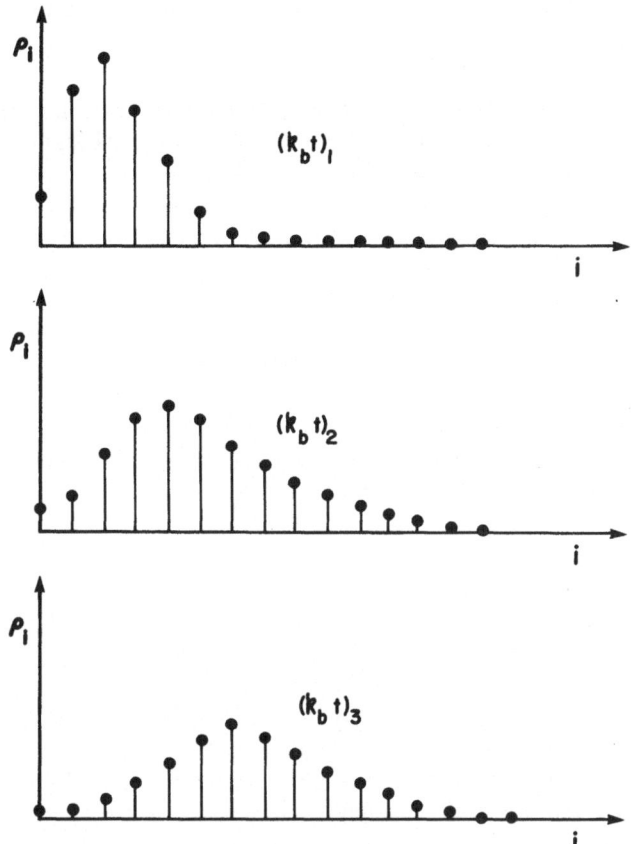

Figure 3.9. The development of the crack size distribution represented by equation (3.14). The represented states are for $(k_b t)_1 = 2.5$; $(k_b t)_2 = 4.5$; and $(k_b t)_3 = 6.5$.

variance, which expresses the spread in crack sizes, is $a^2 k_b t$. These conclusions are particularly significant in cases of slow crack growth, which are of considerable interest for design work and reliability studies.

The sum of two or more Poisson processes also produces a Poisson distribution. Consequently, the sum of parallel mechanisms with $k_b \gg k_h$ (that is, when only breaking activations need be considered) are also described by the normal distribution function. Conversely, probability theory shows that a Poisson process can be separated into its constituent processes. In the analysis of crack growth, this means that rigorous mathematical methods are available to establish the constituent mechanisms of the measured behavior. Discussion of this theory, which describes the so-called renewal processes, is beyond the scope of this chapter. The references [1, 3] serve as particularly useful guides.

One remarkable property of the renewal processes should, however, be noted here: the Poisson crack size and velocity distribution is approximated even by parallel mechanisms that are not of the Poisson type. This theory anticipates that many crack growth processes can be described by the normal probability distribution: the crack sizes at time t are distributed in a normal, Gaussian manner.

Near-threshold conditions: $L_h k_h \simeq L_b k_b$. Many engineering-design decisions hinge on threshold, or near-threshold conditions. Often, when cracks grow very slowly, the breaking and healing rates are nearly equal; at the threshold $L_b k_b = L_h k_h$, and a dynamic equilibrium exists, as discussed in Chapter 2, Part 1. These concepts are, of course, valid only when one energy barrier controls the crack velocity: when multi-barrier kinetics operates, this behavior is conceptually valid. This section presents a method for near-threshold analysis that is simpler and yet more powerful than the general technique presented in Part 1.

Consider two cases of consecutive energy barrier systems. The first is composed of energy barriers for which $L_b k_b = L_h k_h$, while the second is made up of nearly identical barriers with $L_b k_b$ slightly greater than $L_h k_h$. The first condition is, of course, equivalent to the deterministic statement that no net crack growth is observed because, when $L_b k_b = L_h k_h$,

$$v = (L)(k_b - k_h) = 0.$$

The second case expresses the deterministic concept of slow crack growth. The corresponding energy-barrier conditions are represented in Figure 3.10, while Figure 3.11 shows the near-threshold behavior.

Conventional, deterministic considerations expect that no crack growth occurs at $L_b k_b = L_h k_h$, which is, by definition, the threshold. They further anticipate that when $L_b k_b$ is only slightly different from $L_h k_h$, the crack velocity is very small. Probabilistic analyses confirm that these deterministic expectations correspond well enough to the average behavior: however, the

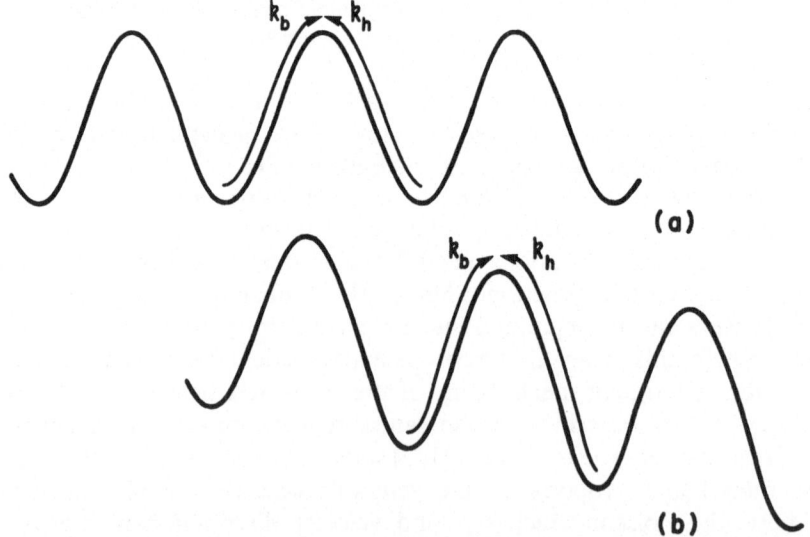

Figure 3.10. The energy-barrier system: (a) equilibrium, threshold, condition, $L_b k_b = L_h k_h$; (b) slow crack growth, $L_b k_b > L_h k_h$.

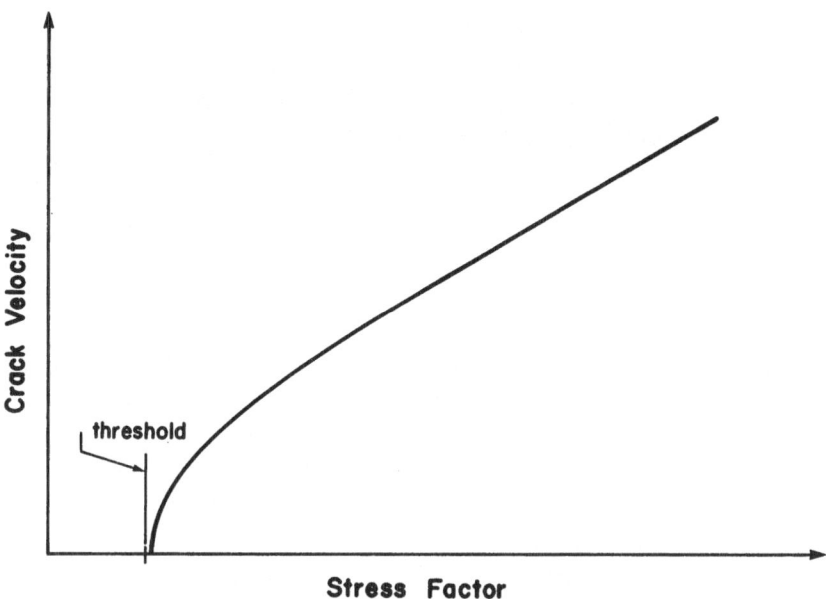

Figure 3.11. The near-threshold behavior.

random nature of the physical process results in a distribution of crack sizes and some cracks may grow quickly.

To describe this behavior, consider that there are ρ_i cracks of size a_i (Figure 3.12), with the crack tips at $\pm x_i$. All breaking rate constants k_b are equal, as are the healing rate constants. Thus the rate of change in the number of cracks, ρ_i, of size a_i is

$$\frac{d\rho_i}{dt} = k_b\rho_{i-1} - (k_b + k_h)\rho_i + k_h\rho_{i+1}. \tag{3.15}$$

The system of differential equations can be replaced by a single partial differential equation, which is equivalent to the Fokker—Planck equation of the random walk theory [4]. This is accomplished by replacing the discrete number of crack sizes, ρ_i, by a continuous distribution of crack sizes, z_i, that represents the crack concentration: the number of cracks per unit length of the crack path. This distribution is the function of size (crack-tip location) and time,

$$z_i(x, t) = \frac{\rho_i}{a},$$

where ρ_i is the number of cracks in the ith energy valley and a is the inter-atomic distance, the distance between two energy valleys. Thus, z_i is the crack size concentration, the number of cracks per unit distance at x_i. Equation (3.15) expresses the continuous distribution of crack sizes as

$$\frac{d z(x, t)}{dt} = k_b z(x - a, t) - (k_b + k_h)z(x, t) + k_h z(x + a, t), \tag{3.16}$$

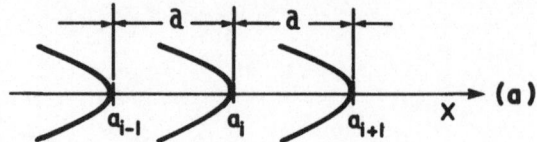

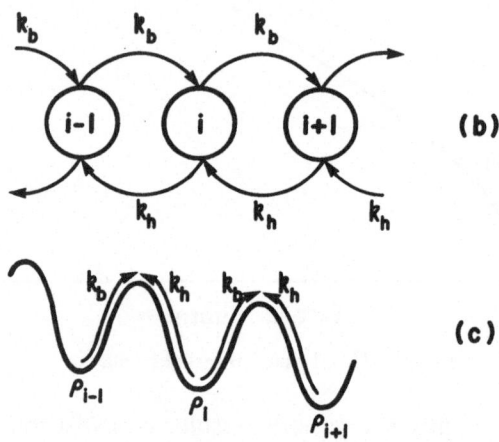

Figure 3.12. (a) The schematic representation of cracks of sizes a_{i-1}, a_i and a_{i+1}; (b) the corresponding Markov-chain model illustrating the transitions among the $i-1$, i and $i+1$ positions; (c) the energy-barrier system: there are ρ_{i-1}, ρ_i and ρ_{i+1} numbers of cracks in the valleys, associated with the corresponding crack sizes.

where $x - a$ signifies that the concentration is considered for cracks with tips at $x - a$, etc. The crack-tip concentrations $\imath(x - a, t)$ and $\imath(x + a, t)$ can be expressed in terms of $\imath(x, t)$ with the Taylor expansion as

$$\imath(x - a, t) \cong \imath(x, t) - \frac{\partial \imath(x, t)}{\partial x} a + \frac{1}{2} \frac{\partial^2 \imath(x, t)}{\partial x^2} a^2, \qquad (3.17)$$

and

$$\imath(x + a, t) \cong \imath(x, t) + \frac{\partial \imath(x, t)}{\partial x} a + \frac{1}{2} \frac{\partial^2 \imath(x, t)}{\partial x^2} a^2. \qquad (3.18)$$

Substitution of equations (3.17) and (3.18) into equation (3.16) results in a partial differential equation [5]

$$\frac{\partial \imath}{\partial t} = \frac{a^2}{2} (k_b + k_h) \frac{\partial^2 \imath}{\partial x^2} - a(k_b - k_h) \frac{\partial \imath}{\partial x}. \qquad (3.19)$$

Equation (3.19) is a typical transport equation, comparable to those for mass transport (diffusion); heat transport (heat flow by conduction); and momentum (viscous fluid) flow. Because all these processes are controlled by the same physical principle, the random fluctuation and flow of energy, they are formulated mathematically in the same way. These processes differ only according to the microscopic conditions of each particular case — the character of the rearrangement of the atomic structure as the energy is redistributed. Similarly, the coefficients of the derivatives are interrelated; the rate constants of breaking and healing are related to the diffusion coefficient, the thermal conductivity and the viscosity coefficient. Further discussion of this interesting topic is beyond the scope of this book. For detailed information, the references can be consulted [6—9].

The solution of equation (3.19), subject to the initial and boundary conditions, describes the distribution of crack sizes with the concentration $c(x, t)$ representing the number of cracks with crack tips at x at time t. The solutions can be developed in functional form using standard mathematical methods, or numerically using the readily available computer programs.

However, there is an important third option that is often more useful and practical. Because equation (3.19) is a typical transport equation, a wide range of solutions are available from the literature of heat and mass transfer [6, 8, 10, 11].

EXAMPLE 3.3. Consider an infinite plate with a through-crack. What is the probability of the crack tip being at x at time t, when initially, at $t = 0$, it was at $x = 0$?

Dividing ρ_i with ρ_t gives the probability distribution. Equation (3.19) is now

$$\frac{\partial P}{\partial t} = \frac{a^2}{2}(k_b + k_h)\frac{\partial^2 P}{\partial x^2} - a(k_b - k_h)\frac{\partial P}{\partial x},$$

where $P(x, t)$ is the probability density: the probability of observing a crack between x and $x + dx$ is $P\,dx$. The solution for the infinite plate is given from the mathematical theory of transport processes. The probability that the crack tip is between x and $x + dx$ is then expressed as [6, 8, 10, 11]

$$P(x, t)\,dx = \frac{1}{[2\pi a^2(k_b + k_h)t]^{1/2}} \exp\left\{-\frac{[x - a(k_b - k_h)t]^2}{2a^2(k_b + k_h)t}\right\}\,dx. \quad (3.20)$$

Figure 3.13 represents the probability density function when shrinkage from the initial size is not restricted. Although this assumption is not physically realistic, it has no macroscopically observable consequences. This is because, as was noted in the Markov-chain analyses of the discrete systems, the normal distribution represented by equation (3.20) develops after a sufficient number of crack growth steps. Because the length of each step is equal to the interatomic distance (approximately 1 to 4 × 10⁻⁷ mm), even a large number of steps, say several hundred, will not essentially alter the initial crack size. Equation (3.20) is thus sufficiently valid. The expectation value

$$E = a(k_b - k_h)t$$

expresses the average crack size. This is, of course, equal to the familiar

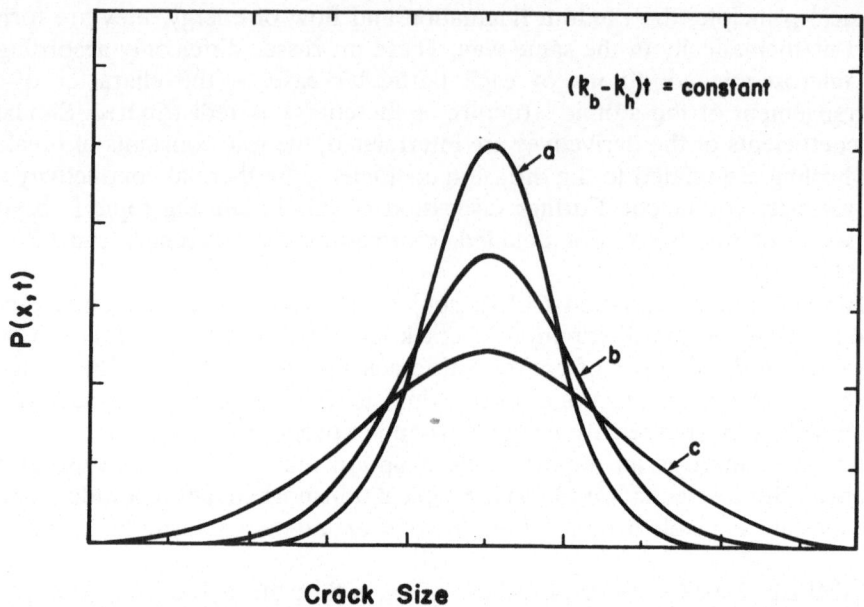

Figure 3.13. The probability density function of crack growth in an infinite plate. For comparison of the variance (the scatter) the distributions are centered: notice the increase in scatter as the function of time. Curve (a) is at $\frac{1}{2}a^2(\mathscr{k}_b + \mathscr{k}_h)t = 1$; (b) at 2; and curve (c) is at 4.

deterministic description of the crack size

$$a = vt = a(\mathscr{k}_b - \mathscr{k}_h)t,$$

where $v = a(\mathscr{k}_b - \mathscr{k}_h)$ expresses the average crack velocity. The spread in crack size distribution is

$$\sigma^2 = a^2(\mathscr{k}_b + \mathscr{k}_h)t,$$

where σ is the standard deviation [5].

EXAMPLE 3.4. The condition $\mathscr{k}_b \cong \mathscr{k}_h$ is particularly important in practical engineering applications. Designs for long service life must ensure very slow crack growth.

In components subjected to stress corrosion cracking, the threshold stress intensity is often regarded as the limit condition at which the crack does not grow. Indeed, it is simple, and usually satisfactory, to treat this as an equilibrium situation where $\mathscr{k}_b = \mathscr{k}_h = \mathscr{k}$. The differential equation then reduces to

$$\frac{\partial P}{\partial t} = a^2 \mathscr{k} \frac{\partial^2 P}{\partial x^2}.$$

The solution is derived easily from transport process mathematics [6, 8, 10, 11]:

$$P(x, t)\, dx = \frac{1}{2(\pi a^2 \ell t)^{1/2}} \exp\left(-\frac{x^2}{4a^2 \ell t}\right) dx.$$

Since the crack cannot shrink back beyond its initial size, the crack distribution is defined as

$$P(x, t)\, dx = \frac{1}{(\pi a^2 \ell t)^{1/2}} \exp\left(-\frac{x^2}{4a^2 \ell t}\right) dx. \tag{3.21}$$

Equation (3.21) expresses the probability of crack sizes between x and $x + dx$ at time t, when the initial crack size, a_0, was small enough to be considered as zero. When the crack was initially of a significantly large half-size then the probability is expressed for crack sizes with respect to $a_0/2$.

Note that $(2a^2 \ell t)^{1/2}$ is equal to the standard deviation: it is a measure of the scatter that results from the physically probabilistic behavior of the crack growth process. Thus, while the deterministic analysis conventionally assumes that no crack growth occurs at the threshold stress intensity, the physically more realistic probability kinetics indicates that there is actually a significant probability of crack growth. Furthermore, once this process begins, crack growth brings about an increasing stress intensity, which in turn accelerates the crack velocity, leading to failure [12].

EXAMPLE 3.5. A condition for very slow crack growth with zero, or near-zero, average velocity is

$$\Delta G_b^+(W) = \Delta G_h^+(W) = \Delta G^+(W)$$

and the elementary rate constants are

$$\ell_b = \frac{kT}{h} \exp\left[-\frac{\Delta G^+(W)}{kT}\right]$$

and

$$\ell_h = \frac{kT}{h} \exp\left[-\frac{\Delta G^+(W)}{kT}\right];$$

hence, $\ell_b = \ell_h = \ell$ and the expectation value is

$$E = a(\ell_b - \ell_h)t = 0;$$

thus the average crack growth velocity is also zero, as required by safety restrictions. But because the process is stochastic, a time-dependent crack size distribution exists, and some cracks will grow with a finite velocity. In other words, there is a finite probability that some components will fail at time t, before the end of the designated service life [12].

EXAMPLE 3.6. The severe consequences of the stochastic character of the thermally activated, time-dependent crack growth process can be illustrated qualitatively by a simple example.

A component must be designed so that when it is subjected to a given stress level, no crack growth will take place in terms of conventional deterministic concepts. For clarity of presentation, the model in this example is an ideally homogeneous material. That is, the component includes a single crack of initial size or, which is the same for the present purpose, a number of identical cracks, each with the same probability of growth. This component is then subjected to a constant W.

The previous example showed that conventional deterministic concepts predict cracks should not propagate when $k_b = k_h$. Thus, to achieve optimum performance, it is desirable to design for a service load that will produce a driving force at which $\Delta G_b^+(W) = \Delta G_h^+(W)$. In this case, $k_b = k_h = k$, the expectation value is

$$a(k_b - k_h)t = 0,$$

and the average growth velocity is zero. Consequently, no crack growth in the conventional sense is expected.

This assumption is dangerous. The discussions in Part 1 showed clearly that because crack growth is a stochastic process, some cracks will grow: a distribution of crack sizes will be observed. Under service conditions, when crack growth can be approximated by the normal distribution function, the standard deviation about the zero average crack velocity is expressed as

$$\sigma = a(k_b + k_h)^{1/2} t^{1/2} = a(2kt)^{1/2}.$$

When

$$\Delta G_b^+(W) = \Delta G_h^+(W) = 0.4 \text{ eV}$$

the standard deviation at 30 years of service life is $\sigma = a(2kt)^{1/2} = 5.35$ mm, and after a service time of six months it is $\sigma = 0.686$ mm.

For a slightly lower apparent activation energy of 0.37 eV, the 30-year standard deviation is nearly 10 mm and is 0.75 mm after only a few months of service. Safety inspections conducted at that time will already reveal cracks in more than one-third of the components. In certain circumstances, fast crack growth can take place even when the cracks are smaller than the critical size. Thus, the weakened components may need replacement. These considerations are particularly significant for ceramic materials.

It is crucial to recognize, therefore, that components such as pipes, pressure vessels and containers considered to be safe in conventional design may fail under the service conditions discussed above.

3.4. The random walk concept of crack growth

In Part 1 of this chapter it was shown that the probabilistic character of the frequency of bond breaking and healing steps identifies crack growth with random walk, or Brownian movement, processes. The following discussion develops the typical mathematical tools of crack growth analysis as a random walk process. In reviewing this example, the conceptual correlation of the probabilistic behavior with the Markov-chain mathematical technique should be kept in mind.

While the mathematical formulation of the random walk theory is rigorous, it has not yet been developed into a practical alternative for engineering applications. This section should therefore be read for the insight it provides into the physical behavior of crack growth, and for opportunities to correlate the crack growth mechanism with other random walk, Brownian movement processes.

EXAMPLE 3.7. Consider that in one-dimensional crack movement the tip moves forward or backward along the x-axis. The crack-tip movement is random: that is, either bond breaking or bond healing can take place with a probability of p_b or p_h respectively, so that $p_b + p_h = 1$. Consider now that N_b bond-breaking and N_h bond-healing activations occur and that after $N_b + N_h = N$ activations the crack tip is at position m along the x-axis.* The breaking- and healing-steps are completely at random and each results from uncoordinated, independent activation. A specific sequence of breaking- and healing-steps is defined by the following set [11, 13, 14]:

$$b\,b\,h\,b\,b\,b\,h\,h\,b\,b \ldots ,$$

and, from this set, the probability that the crack tip reaches m is

$$p_b p_b p_h p_b p_b p_b p_h p_h p_b p_b \cdots$$

$$p_b p_b p_b p_b p_b p_b p_b \cdots p_h p_h p_h \cdots = p_b^{N_b} p_h^{N_h}$$

Although the probabilities of the steps are multiplied, the arrangement of the two different step types within the sequences must be considered. For N_b breaking and N_h healing steps, the number of sequences that may reach m without repeating an arrangement is $N!/N_b!N_h!$. Thus, the probability $P(m, N)$ that the crack tip reaches the point m after N steps can be formulated [9] as follows:

$P(m, N) \equiv$ probability of any specific sequence reaching m after N steps, times the number of the distinct sequences that reach m, or:

$$P(m, N) = p_b^{N_b} p_b^{N_h} \frac{N!}{N_b!N_h!} . \tag{3.22}$$

---

* Note that m is expressed in units of the interatomic distance a.

The probability of a bond-breaking step p_b or a bond-healing step p_h is given by

$$p_b = \frac{\text{rate of bond-breaking activations}}{\text{rate of bond-breaking + bond-healing activations}}$$

$$= \frac{k_b}{k_b + k_h}$$

and   (3.23)

$$p_h = \frac{\text{rate of bond-healing activations}}{\text{rate of bond-breaking + bond-healing activations}}$$

$$= \frac{k_h}{k_b + k_h},$$

where k_b and k_h express the average number of breaking- and healing-activations per unit time over the energy barriers of the thermally activated process.

Because $N = N_b + N_h$ and $m = N_b - N_h$, it follows that

$$N_b = \tfrac{1}{2}(N + m)$$

$$N_h = \tfrac{1}{2}(N - m).$$ (3.24)

Combining equations (3.23) and (3.24), equation (3.22) becomes

$$P(m, N) = \left(\frac{k_b}{k_b + k_h}\right)^{\frac{1}{2}(N + m)} \left(\frac{k_h}{k_b + k_h}\right)^{\frac{1}{2}(N - m)} \frac{N!}{[\frac{1}{2}(N + m)]! [\frac{1}{2}(N - m)]!}. \quad (3.25)$$

Anticipating the mathematical development, the introduction of the symbols $k^+ = k_b + k_h$ and $k^- = k_b - k_h$ will result in

$$k_b = \tfrac{1}{2}(k^+ + k^-),$$

$$k_h = \tfrac{1}{2}(k^+ - k^-),$$ (3.26)

and the substitution of equations (3.26) into (3.25) expresses $P(m, N)$ as

$$P(m, N) = [\tfrac{1}{2}(1 + k_p)]^{\frac{1}{2}(N + m)} [\tfrac{1}{2}(1 - k_p)]^{\frac{1}{2}(N - m)} \frac{N!}{[\frac{1}{2}(N + m)]! [\frac{1}{2}(N - m)]!} \quad (3.27)$$

where $k_p = k^- / k^+$.

In the asymptotic series $N!$ is

$$N! \simeq (2\pi N)^{1/2} \left(\frac{N}{e}\right)^N \left(1 + \frac{1}{12} N + \frac{1}{288} N^2 - \frac{139}{51,840} N^3 - \cdots\right),$$

and even if $N = 10$, the error is just 0.8% when only the terms of the series from Sterling's formula

$$N! \simeq (2\pi N)^{1/2} \left(\frac{N}{e} \right)^N$$

are used.

Introducing then the approximation and taking the logarithm of equation (3.27), the probability expression becomes

$$\ln P(m, N) = \tfrac{1}{2}(N + m) \ln \tfrac{1}{2} + \tfrac{1}{2}(N + m) \ln(1 + \ell_p) + \tfrac{1}{2}(N - m) \ln \tfrac{1}{2} +$$
$$+ \tfrac{1}{2}(N - m) \ln(1 - \ell_p) - \tfrac{1}{2} \ln 2\pi + (N + \tfrac{1}{2}) \ln N -$$
$$- \tfrac{1}{2}(N + m + 1) \ln \frac{N}{2} - \tfrac{1}{2}(N + m + 1) \ln \left(1 + \frac{m}{N} \right) -$$
$$- \tfrac{1}{2}(N - m + 1) \ln \frac{N}{2} - \tfrac{1}{2}(N - m + 1) \ln \left(1 - \frac{m}{N} \right).$$

Because ℓ_p and m/N are less than 1, the binomial theorem

$$\ln(1 \pm a) = \pm a - \frac{a^2}{2} \pm \frac{a^3}{3} - \cdots$$

may be applied, and therefore

$$\ln P(m, N) = \tfrac{1}{2}(N + m) \ln \tfrac{1}{2} + \tfrac{1}{2}(N + m) \left(\ell_p - \frac{\ell_p^2}{2} \right) + \tfrac{1}{2}(N - m) \ln \tfrac{1}{2} +$$

$$+ \tfrac{1}{2}(N - m) \left(-\ell_p - \frac{\ell_p^2}{2} \right) - \tfrac{1}{2} \ln 2\pi + (N + \tfrac{1}{2}) \ln N -$$

$$- \tfrac{1}{2}(N + m + 1) \ln \frac{N}{2} - \tfrac{1}{2}(N + m + 1) \left(\frac{m}{N} - \frac{m^2}{2N^2} \right) -$$

$$- \tfrac{1}{2}(N - m + 1) \ln \frac{N}{2} - \tfrac{1}{2}(N - m + 1) \left(-\frac{m}{N} - \frac{m^2}{2N^2} \right)$$

$$= \ln 2 - \tfrac{1}{2} \ln(2\pi N) + m\ell_p - N \frac{\ell_p^2}{2} - \frac{m^2}{2N} + \frac{m^2}{2N^2}. \qquad (3.28)$$

In equation (3.28), terms of the order $a^3/3$ and higher have been neglected. When, however, the value of a^3 is not much less than a^2, additional terms of the expansion series must be considered and will appear in the probability expression.

In can be shown that a further approximation by omitting $m^2/2N^2$ from equation (3.28) produces about 10% error in the probability distribution when $m/N = 0.5$, and when $m/N = 0.1$, the error is less than 0.5%. Consequently, the probability of finding the crack tip at m after N steps is found to be

$$P(m, N) = \frac{2}{(2\pi N)^{1/2}} \exp\left[-\frac{(m - N\ell_p)^2}{2N}\right]. \tag{3.29}$$

Equation (3.29) represents a discrete, symmetrical, bell-shaped probability distribution function. To represent the probability as a continuous function, $P(m, N)$ must be expressed in terms of the net displacement x and time t.

The net displacement of the crack tip is

$$x = mL,$$

where L is the *average distance* after each activation, that is, the interatomic distance or its integer multiple. It should be noted here that for even numbers of N, m must be even, and if N is odd, m must also be odd. Consequently, the interval Δx along the path is

$$\Delta x = 2\Delta mL$$

and

$$P(x, N)\Delta x = P(m, N)\left(\frac{\Delta x}{2L}\right) = \left(\frac{1}{2\pi L^2 N}\right)^{1/2} \exp\left[-\frac{(x - LN\ell_p)^2}{2N}\right]\Delta x. \tag{3.30}$$

The probability distribution curve expressed by equation (3.30) is symmetrical about the point $\bar{x}$, the average distance travelled by the crack tip. The average distance is given by

$\bar{x} \equiv$ (average number of steps forward $-$ average number of steps
      backward) $\times$ step length

$$= L\ell_b t - L\ell_h t = L(\ell_b - \ell_h)t = L\ell^- t.$$

Since equation (3.30) reaches its maximum value when the exponent is zero — that is, when $x = \bar{x}$ — so

$$\bar{x} - LN\ell_p = 0,$$

or

$$L\ell^- t - LN\left(\frac{\ell^-}{\ell^+}\right) = 0.$$

It follows that

$$N = \ell^+ t. \tag{3.31}$$

Finally, substituting equation (3.31) into (3.30), the probability $P(x, t)\Delta t$ that the crack tip can be found in the interval x and $x + \Delta x$ at time t is

$$P(x, t)\,\Delta x = \left(\frac{1}{2\pi L^2 \ell^+ t}\right)^{1/2} \exp\left[-\frac{(x - L\ell^- t)^2}{2L^2 \ell^+ t}\right]\Delta x. \qquad (3.32)$$

In the special case when $\ell_{\mathrm{b}} = \ell_{\mathrm{h}} = \ell$, equation (3.25) reduces to

$$P(m, N) = (\tfrac{1}{2})^N \frac{N!}{[\tfrac{1}{2}(N + m)]!\,[\tfrac{1}{2}(N - m)]!}, \qquad (3.33)$$

because

$$\tfrac{1}{2}(N + m) + \tfrac{1}{2}(N - m) = N.$$

Following the arguments leading to equations (3.29) and (3.31) it can be shown readily that

$$P(m, N) = \left(\frac{2}{\pi N}\right)^{1/2} \exp\left(-\frac{m^2}{2N}\right),$$

and

$$P(x, t)\,\Delta x = \frac{1}{2(\pi L^2 \ell t)^{1/2}} \exp\left(-\frac{x^2}{4L^2 \ell t}\right)\Delta x, \qquad (3.34)$$

since $N = 2\ell t$.

While equations (3.27) and (3.33) are exact solutions of the crack growth process, they are inconvenient to evaluate. Equations (3.32) and (3.34), on the other hand, are essentially identical with the solution of the differential equation of crack growth demonstrated in Section 3.3.

When the probability distribution function of equation (3.32) is represented in the $P(x, t)$ versus x coordinate system, $\ell^+ t$ determines the shape of the curves, while $\ell^- t$ defines the location of its maximum. Figure 3.13 illustrates the variations of the probability curves for different values of $\ell^+ t$. Figure 3.14(a) represents the shift of the maximum along the x-axis, as well as the positions and sense of movement of the crack tips. The corresponding energy-barrier systems of the thermally activated crack growth processes are shown schematically in Figure 3.14(b).

The study of crack growth as a random walk phenomenon provides unique insight into the essence of the process. It demonstrates that the measured scatter in any set of data is the sum of several factors: the statistical crack size distribution resulting from the heterogeneity of the material; the scatter of experimental or service conditions; and the diffuse character of the physical process itself, which is basically a random walk phenomenon.

In principle, the first two factors can be influenced by the experimentalist and

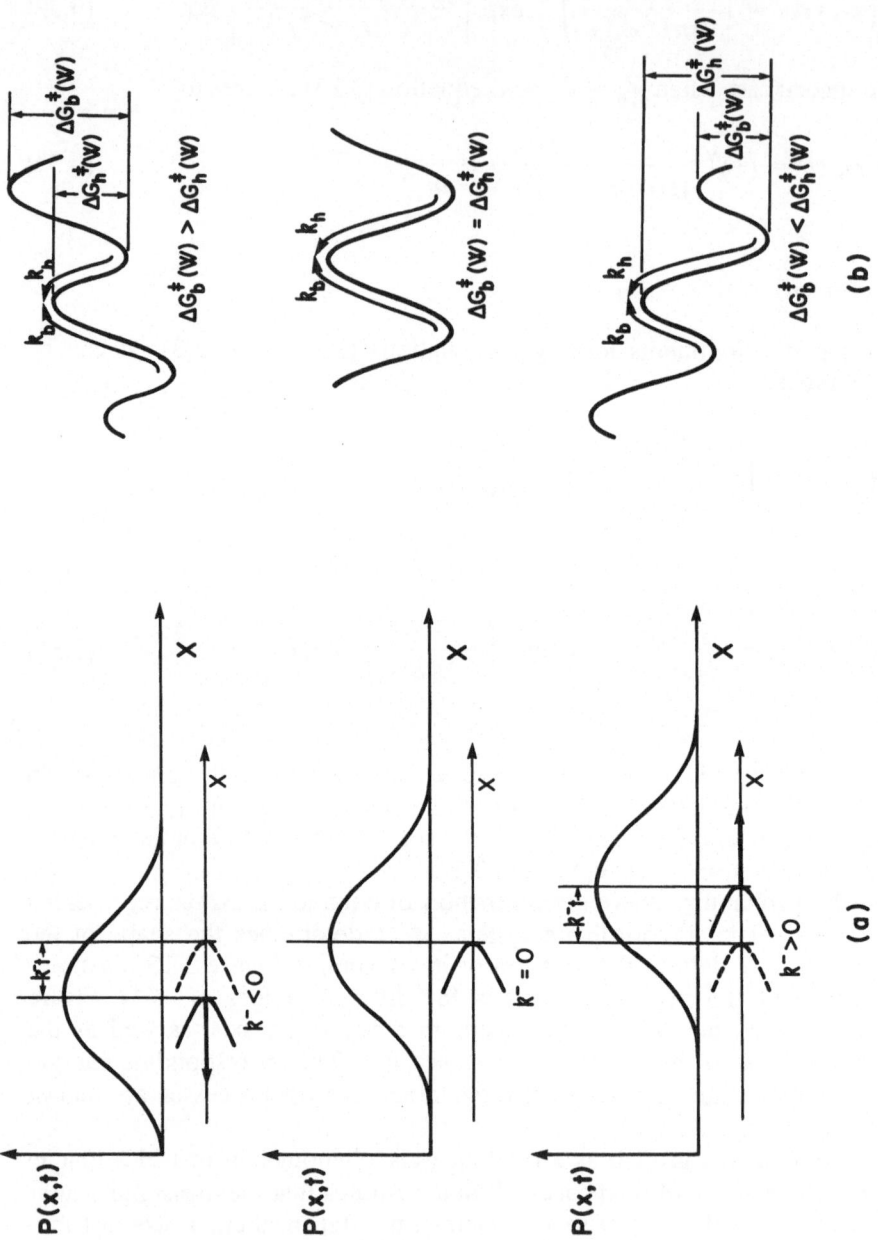

Figure 3.14. (a) Illustration of the shift in the location of the probability curve maximum, and the direction of crack tip movement, as influenced by the value of $\dot{\mathscr{k}}^- t$ in equation (3.32); (b) schematic representation of the corresponding energy barrier systems. The average crack size decreases when $\Delta G_b^{\ddagger}(W) > \Delta G_h^{\ddagger}(W)$, is stationary when $\Delta G_b^{\ddagger}(W) = \Delta G_h^{\ddagger}(W)$, and increases when $\Delta G_b^{\ddagger}(W) < \Delta G_h^{\ddagger}(W)$.

by the design or operating engineer. However, the random walk effect is always present, even under ideal conditions, and its magnitude depends on the value of $\mathscr{k}^+t$, that is, the term in equation (3.32) that determines the shape of the crack-size-probability distribution function. The diffusive nature of the process, therefore, can be controlled through ΔG^+, W, T and time. The subtraction of the physical scatter from the measured scatter indicates the means by which testing and service performance can be improved. The lower limit of scatter, below which testing or service conditions cannot be reduced, is fixed by the physical scatter that results from the random walk character of the process. This minimum scatter, which is inevitably present in the service life of structures and machine components, must be considered statistically as the limiting condition for design lifetime.

It is significant that the mathematical treatment of crack velocity can draw on the wealth of analytical results that have been developed for other random walk processes.

3.5. Stress corrosion cracking

While for many purposes the deterministic model of SCC described in Chapter 2, Part 2, provides a good approximation, the probabilistic, discrete nature of bond-breaking and healing steps is more accurately represented by the Markov-chain model. The corresponding constitutive equations were developed in Part 1 of this chapter. It was noted there that the mathematical processing of the system can be simplified by taking advantage of the specific conditions of the process. This section shows how to do this, while extending previous descriptions of SCC [15—17].

It was shown in Chapter 2 that Regions I and II of Figure 3.15 are controlled by a mechanism that corresponds to a system of consecutive energy barriers. The corrosive environment provides the reagent molecules and delivers them to the crack tip, where a chemical reaction changes the matrix into a usually brittle, weakened material. In this process bonds are then broken by the combined action of mechanical work and the random, thermal fluctuations of the atoms at the crack tip. The diffusion, chemical reaction and bond-breaking steps are thermally activated, and because diffusion and chemical reaction must take place for the bonds to break, the process is sequential over the stress factor range of Regions I and II.

In each application the analysis has to be carried out for the actual mechanism. For instance (see Chapter 2), in Region I chemically enhanced bond breaking and in Region II reactant transport is the typical rate-controlling process for ceramics. For clarity and expediency of the representation of the complex kinetics and for ease of visualization, the example includes only two of the processes: the sequential steps of chemical reaction and bond-breaking mechanism, with simplified notation.

The atomic process is shown schematically in Figure 3.16, together with the associated energy-barrier system. The location of the crack tip in the pre-

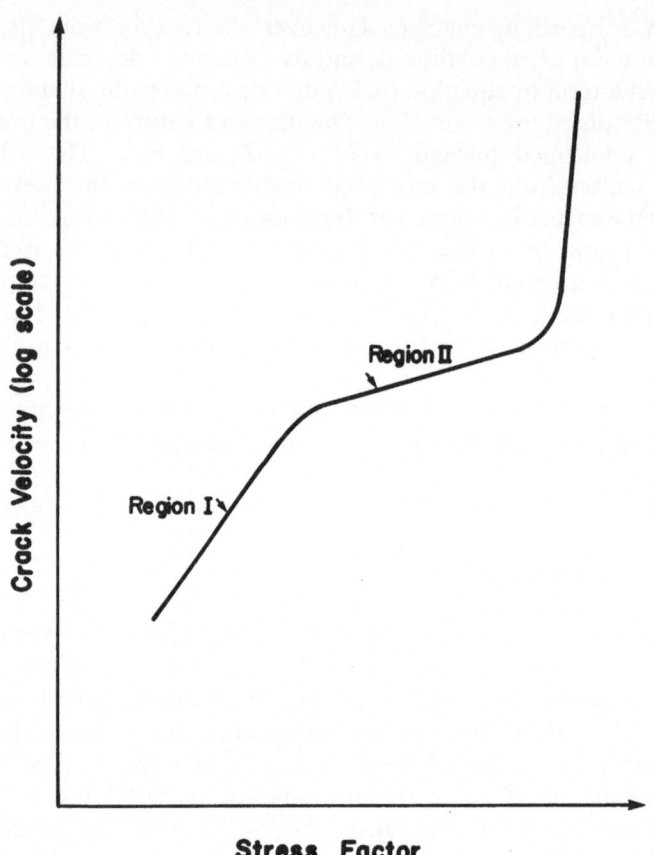

Figure 3.15. Schematic diagram of the typical stress corrosion cracking behavior.

cracked specimen is at x_0, Figure 3.16(a). The crack cannot move below this size, that is, $\mathscr{k}_h = 0$ for the initial stage, because the energy goes to infinity in front of valley 0, Figure 3.16(c). As the specimen is immersed in the reactive environment, a corrosion reaction takes place, Figure 3.16(b), and the system moves over the first barrier and down to valley 1. For mathematical convenience here, the corrosion reaction is considered to be irreversable, that is, the chemical dissolution rate constant is zero. While this is physically reasonable, it is by no means a general and necessary assumption and should not be taken as a restriction on the validity of the theory. The load is applied simultaneously with the immersion and the next step, the breaking of the bond, moves the tip by one atomic distance into valley 2 at $x_0 + a$. The freshly exposed surface is now ready for the repetition of the two steps: each cycle consists of two barriers. Figure 3.16(c) illustrates the energy-barrier system of the process and identifies the subscripts of equation (3.35), while Figure 3.16(d) is the typical Markovian model.

The crack growth process is now fully identified and the rate equation system can be formulated. The description of the model is defined as

$$\frac{d\rho_0}{dt} = -k_c\rho_0 \qquad\qquad \text{initial stage}$$

$$\left.\begin{aligned}
\frac{d\rho_1}{dt} &= k_c\rho_0 - k_b\rho_1 + k_h\rho_2 \\[2ex]
\frac{d\rho_2}{dt} &= k_b\rho_1 - (k_h + k_c)\rho_2
\end{aligned}\right\} \qquad \text{first cycle}$$

$$\left.\begin{aligned}
\frac{d\rho_3}{dt} &= k_c\rho_2 - k_b\rho_3 + k_h\rho_4 \\[2ex]
\frac{d\rho_4}{dt} &= k_b\rho_3 - (k_h + k_c)\rho_4
\end{aligned}\right\} \qquad \text{second cycle} \qquad (3.35)$$

$$\vdots$$

$$\left.\begin{aligned}
\frac{d\rho_{n-2}}{dt} &= k_c\rho_{n-3} - k_b\rho_{n-2} + k_h\rho_{n-1} \\[2ex]
\frac{d\rho_{n-1}}{dt} &= k_b\rho_{n-2} - (k_h + k_c)\rho_{n-1}
\end{aligned}\right\} \qquad \text{last cycle}$$

$$\frac{d\rho_n}{dt} = k_c\rho_{n-1} - k_{bf}\rho_n \qquad\qquad \text{final stage}$$

Beginning from the second equation in this model, the rate constants of the odd-numbered concentration-changes ($d\rho_1/dt$, $d\rho_3/dt$, ...) are equal, and a similar observation can be made of the even-numbered concentration-change equations. It follows that the form of the equations describing the concentration changes in the corrosion and bond-breaking stages repeats with each cycle, as indicated by the parentheses of equations (3.35). Collecting the odd terms from each cycle leads to

$$\frac{d\rho_{odd}}{dt} = k_c\rho_{even} - k_b\rho_{odd} + k_h\rho_{even}$$

where ρ_{odd} (breaking step, stage 1) represents the summation of the number of crack tips in the odd-numbered valleys from 1 to $n-1$. Since the final stage is distinct from the cycles, the summations include only the last full cycle. Similarly, the collection of the even-numbered concentration changes for the full cycles leads to the expression

$$\frac{d\rho_{even}}{dt} = k_b\rho_{odd} - (k_h + k_c)\rho_{even}.$$

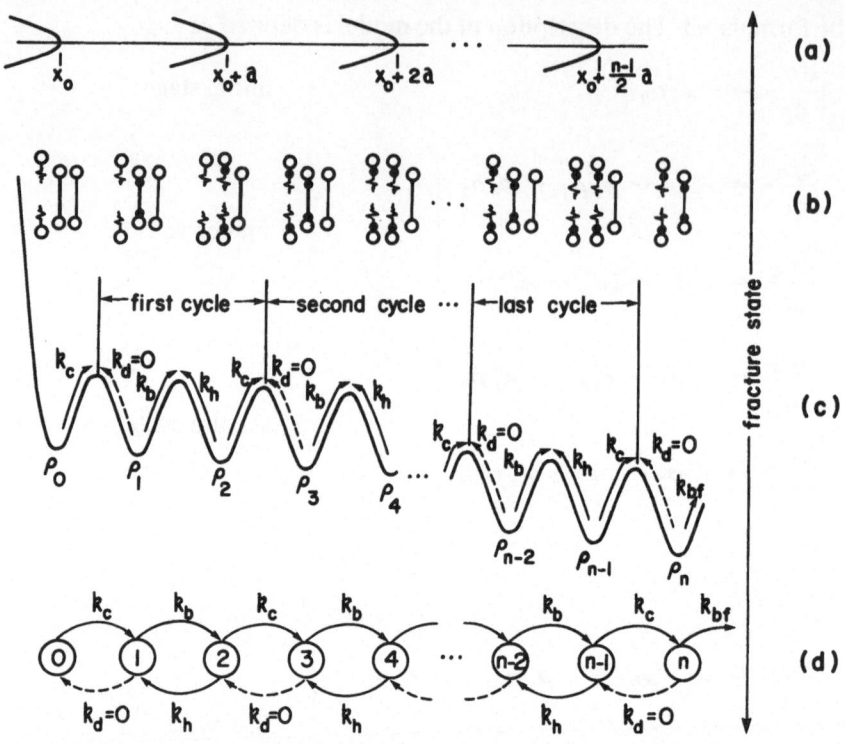

Figure 3.16. The schematic representations of two steps of crack growth in SCC. (a) The consecutive steps in crack growth from the initial size x_0 through $x_0 + a$, $x_0 + 2a$, to fracture. (b) The schematic of the atomic-level occurrences. The open circles represent the atoms of the component's material; the solid circles represent the ions of the degrading material; the bars indicate intact bonds, and the broken lines indicate broken bonds. (c) The energy-barrier system with the associated rate constants and their direction of effect identified. The number of cracks in the valleys are denoted by ρ. (d) The typical Markov-chain model.

The introduction of these into equations (3.35) reduces the system to four equations:

$$\frac{d\rho_0}{dt} = -k_c \rho_0$$

$$\frac{d\rho_{\text{odd}}}{dt} = -k_b \rho_{\text{odd}} + (k_c + k_h)\rho_{\text{even}}$$

$$\frac{d\rho_{\text{even}}}{dt} = k_b \rho_{\text{odd}} - (k_h + k_c)\rho_{\text{even}}$$

(3.36)

$$\frac{d\rho_n}{dt} = k_c \rho_{n-1} - k_{bf} \rho_n.$$

One condition underlying equations (3.36) is that the stress intensity factor

K, and thus the work, are constant over the entire range of crack growth. The rate constants ℓ are therefore independent from the position of the crack. Under experimental conditions this is usually achieved by using appropriately shaped specimens. When crack growth is restricted to a distance far from the end edge, the probability of finding the crack tip near to n is very small and thus $\rho_n \simeq 0$. Then $d\rho_n/dt = 0$ and equations (3.36) are limited to the three expressions of $(d\rho_0/dt)$, $(d\rho_{odd}/dt)$ and $(d\rho_{even}/dt)$.

The three remaining stages in equation (3.36) are solved by using equation (3.5) of Part 1. The full Markov-chain development of the solution is presented in Appendix D. Although this is admittedly laborious, it follows a clearly established procedure. The appendix might require specialized interest, but it should provide insight into the full mathematical sequence. Longer or more complex Markov-chain representations of probabilistic crack growth processes involve identical steps — the additional work is required only to expand the scope. For instance, while the secular equation for a larger determinant obviously requires greater time and effort, no new concepts or mathematical techniques are needed. Computer programs are available as standard subroutines.

3.6. Corrosion fatigue

Corrosion fatigue, a major cause of failure, results from the simultaneous action of a fluctuating mechanical load and the degradation of the material in the crack-tip zone. Conventional analyses of corrosion fatigue behavior recognize that the crack growth rate da/dN, where N is the number of cycles depends on, among others, the duration and shape of the cycle and the temperature [18—23].

In the description of deterministic fracture kinetics of corrosion fatigue it was noted that the rate-controlling process is thermally activated and is thus defined by fracture kinetics. Application of the theory led to fracture constitutive laws that are essential in understanding the crack growth process for engineering applications. However, it should be emphasized again that the validity of deterministic models of corrosion fatigue are more restricted than probabilistic formulations and cannot provide information on effects that result from the random character of the physical process of thermal activation.

One consequence of this can be anticipated by considering the mechanism of corrosion fatigue. Its rate-controlling mechanism includes two basic processes: transport of the reactant to the crack-tip zone and chemical bond weakening. Both consist of several elementary steps. However, to facilitate understanding of the kinetic concepts and the mathematical description, the following application is restricted to a corrosion step represented by a single rate constant, ℓ_c, and to crack propagation steps controlled by bond breaking and healing, represented by the ℓ_b and ℓ_h rate constants, respectively, as discussed in Section 3.5 [22—24].

The elementary steps of corrosion and the subsequent crack growth take place in sequence. A certain proportion of cracks is in either of the two principal states: that is, some cracks are ready to corrode, while others are ready for a bond-breaking or healing step. The distribution of the cracks in the two states is probabilistic and, of course, the greater the proportion of cracks with tip zones that are already corroded, the faster the growth rate. This means that not only the frequency and temperature are important, as recognized by deterministic kinetics, but also the distribution of the cracks between the two principal states; and this is a probability concept [25].

This distribution depends on the rate constants k_c, k_b and k_h. The previous section showed that in corrosion-affected crack growth (step function loading) the distribution initially undergoes a transition phase, when the fraction of cracks in the corrosion process state is ρ_{odd}/ρ_t, while in the growth process state is ρ_{even}/ρ_t. Both fractions change over time, as expressed by the relations

$$\frac{\rho_{odd}}{\rho_t} = F \exp(-\lambda_2 t) + G \exp(-\lambda_3 t) + H$$

and

$$\frac{\rho_{even}}{\rho_t} = I \exp(-\lambda_2 t) - G \exp(-\lambda_3 t) + M,$$

where *F, G, H, I* and *M* are functions of the rate constants k_c, k_b, k_h, as derived in detail in Appendix D [26].

It is expected that when this type of analysis is extended to describe the effects of waves with various shapes and frequencies, it will advance considerably the understanding of corrosion fatigue. However, this development lies in the future.

The next example illustrates the effect for a simple case that follows from the above equations.

EXAMPLE 3.8. Crack distribution curves were calculated with equations (D.7) of Appendix D for a hypothetical material, tested in toluene. As Figure 3.17 shows, the transition period for loads characterized by work values of less than $\sim 0.77 \, eV$ was greater than 1×10^{-2} s. This, of course, pertains to the single ramp-type loading condition. But it is informative enough to illustrate that, for the discussed environment, cycle frequencies of higher than 50 Hz are subject to the redistribution effect [6, 26, 27, 28].

The analysis demonstrates that there is a time lag between the crack velocity and the rate of stress change. It follows from the logic of the mechanism that under increasing loading, the transition effect retards propagation; while under decreasing loads the growth is faster. Consequently, when the half-cycle period (considered in terms of the effective time of crack closure) is shorter than the transition period, the crack growth behavior is substantially affected by the non-steady state of the system.

3.7. Composite materials

High-strength composite laminates are important engineering materials. Their

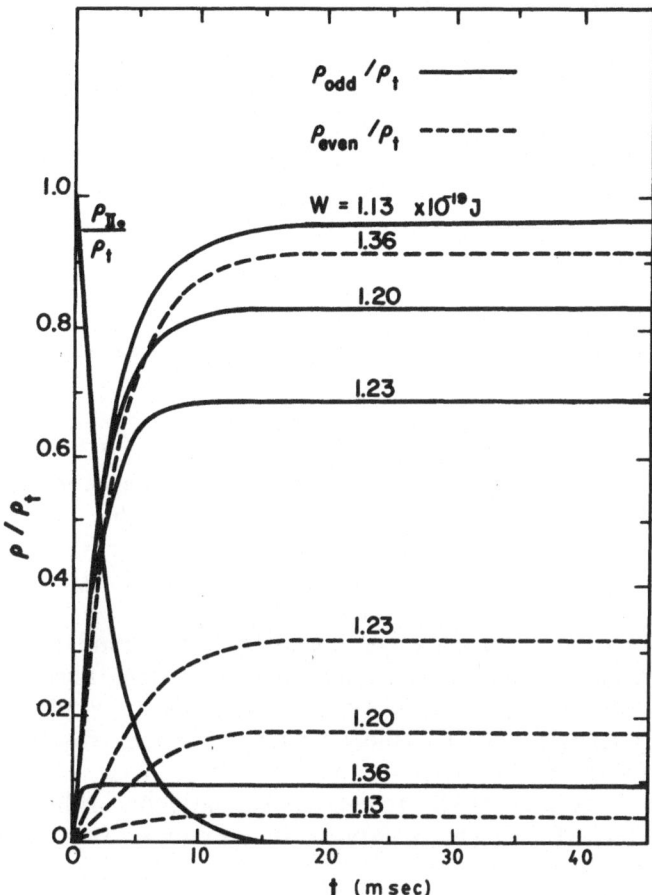

Figure 3.17. Variation of the number of crack tips in front of the two cyclically repeating barriers. The curves were calculated by using equations (D.7) of Appendix D, for four different W values. The other parameters are $\Delta G^+ = 2.24 \times 10^{-19}$ J; $T = 300$ K; $\ell_c = 350$ s^{-1}.

industrial use presupposes an understanding of crack growth across alternating hard and soft layers. Furthermore, many composites are very sensitive to temperature and environment, and these effects demand special attention [28–30].

The layers of the composites will be designated in the discussion as either hard or soft. It should be noted that the terms low strength and high strength may be substituted, but that the two sets of designations are not necessarily interchangeable. Accordingly, the rate constants in the hard layer are expressed as

$$\ell_b^h = \frac{kT}{h} \exp\left(-\frac{\Delta G_b^{+h} - W_b^h}{kT}\right) \quad \text{and} \quad \ell_h^h = \frac{kT}{h} \exp\left(-\frac{\Delta G_h^{+h} + W_h^h}{kT}\right);$$

and in the soft layer as

$$\ell_b^s = \frac{kT}{h} \exp\left(-\frac{\Delta G_b^{+s} - W_b^s}{kT}\right) \quad \text{and} \quad \ell_h^s = \frac{kT}{h} \exp\left(-\frac{\Delta G_h^{+s} + W_h^s}{kT}\right).$$

Close attention to the previous text is now rewarded by a wider and deeper appreciation of the fracture process: the Markov chain can be written in terms of the probability of finding a crack growth to size a_i, defined as

$$P_i = \frac{\rho_i}{\rho_t}.$$

Probability terminology expresses the general term of the Markov chain as

$$\frac{dP_i}{dt} = \ell_b P_{i-1} - (\ell_b + \ell_h)P_i + \ell_h P_{i+1},$$

where for the ℓ_b and ℓ_h activation rate constants the appropriate expression for the hard or soft layer is chosen, depending on whether the layer is in the $i-1$, i, or $i+1$ state. Extension of the description to rate constants that depend on the crack size follows from the previous discussions.

EXAMPLE 3.9. To determine the crack-size distribution in layered composites, the Markov-chain differential equation system must be evaluated numerically. The strength of the layers is expressed by the atomic bond strengths: accordingly, the activation energy of crack growth ΔG^+ varies periodically. Figure 3.18(a) illustrates the alternation of high- and low-strength layers, while Figure 3.18(b) illustrates a typical variation of the corresponding activation energy, ΔG^+. Figure 3.19 presents an example. The activation energy

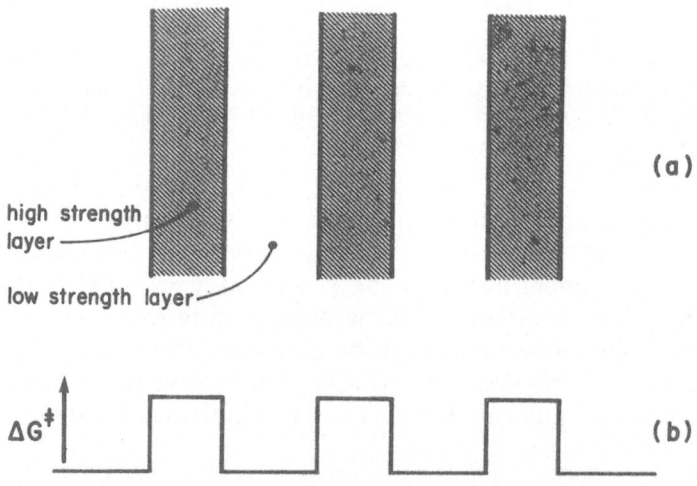

Figure 3.18. (a) Schematic representation of the composite material; and (b) the corresponding variation of the activation energy.

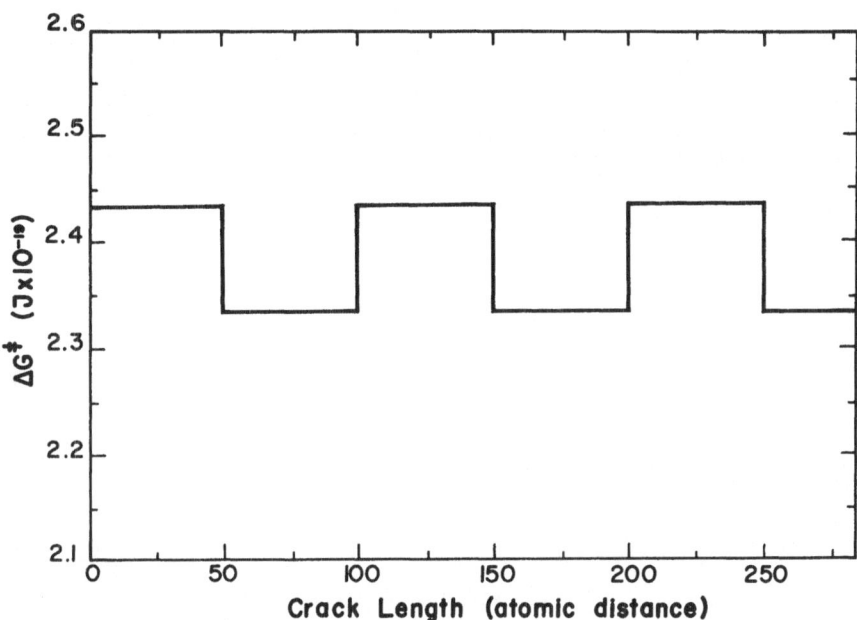

Figure 3.19. (a) The distribution of the activation energy discussed in Example 3.9.

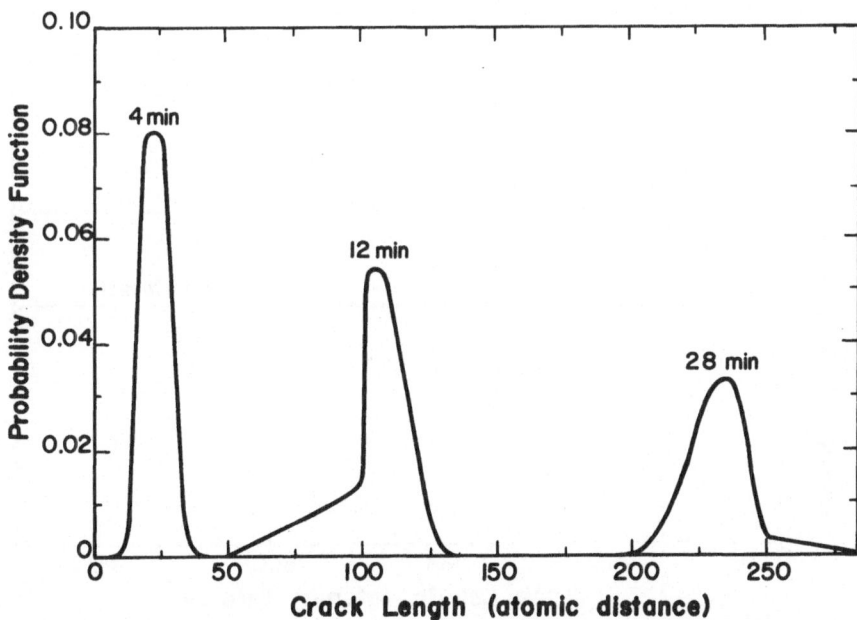

Figure 3.19. (b) The calculated crack-size distribution. The square wave strength variation was represented; $\Delta G_b^{+h} = 2.43 \times 10^{-19}$ J, $\Delta G_b^{+s} = 2.33 \times 10^{-19}$ J, $W_b^h = W_b^s = 1.28 \times 10^{-19}$ J, $\Delta G_h^{+h} = \Delta G_h^{+s} = 0$, $T = 300$ K.

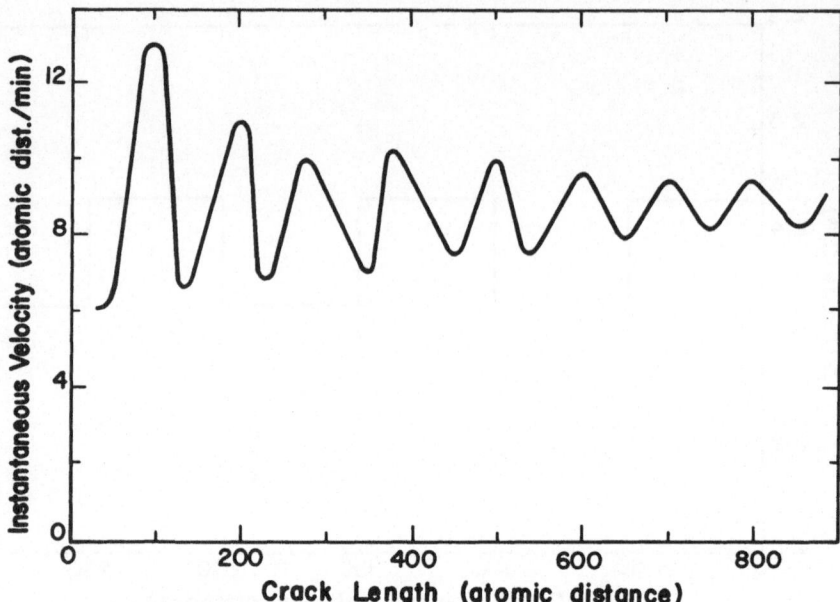

Figure 3.19. (c) The instantaneous crack velocity.

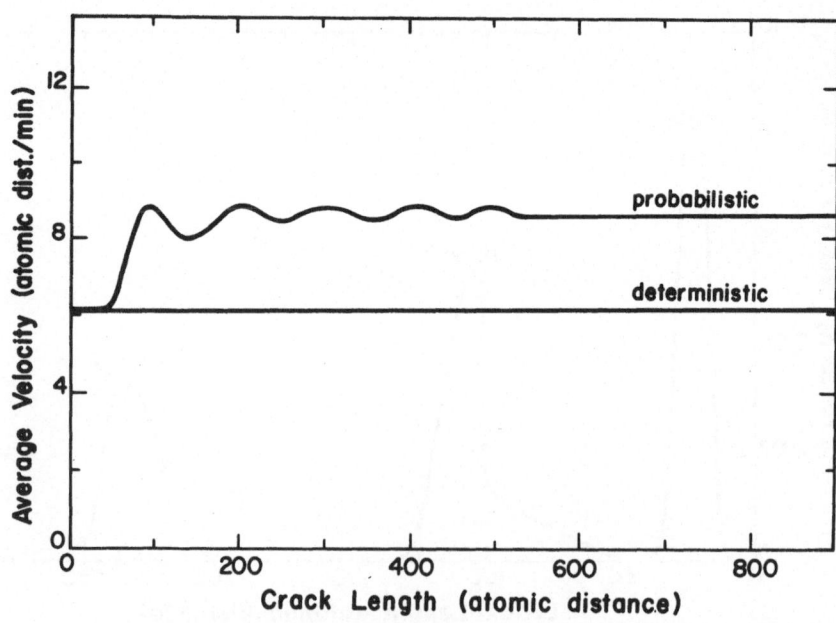

Figure 3.19. (d) Comparison of the probabilistic and deterministic crack velocities as calculated in Example 3.9.

varies as shown in Figure 3.19(a). Over the time period, illustrated in Figure 3.19(b), crack sizes vary by several layers. Figure 3.19(c) illustrates that the instantaneous crack velocity fluctuates initially, but steadies to a probabilistic value that is significantly higher than that predicted by the deterministic model, shown in Figure 3.19(d).

In this study [29], computer capacity restricted the maximum number of crack growth steps to about 1000. To obtain realistic information from this relatively small sample, layer thicknesses were given only in terms of a finely layered, lamellar, eutectic material — a natural composite. The results, however, are qualitatively valid for macroscopic, artificial layered composites as well.

Composite materials are being used at an increasing rate and are attracting much interest in engineering research, design and manufacturing [30—32]. It was pointed out that crack growth analysis of composite materials is a difficult undertaking. The heterogeneity of the material, and the large variety of crack growth mechanisms that are sensitive to time, temperature and environment render the lifetime determination 'a formidable challenge' [30]. In this task, fracture kinetics is a powerful tool [33—35]. It is to be noted that alternative considerations of structural randomness also lead to fracture kinetics-type analyses [36—39].

3.8. Comments and summary

Part 2 of this chapter explored the concepts and mathematical apparatus of probabilistic fracture kinetics using a wide range of examples, and illustrated correlations of probabilistic kinetics with the theory of random walk, or Brownian motion. The discussion showed that because probabilistic kinetics analyses express the probabilistic aspect of crack growth, they provide descriptions that are essential for effective, safe design. This is particularly true in the following cases.

(1) In near-threshold conditions the breaking and healing activation rates may be almost equal. Consequently a Gaussian distribution of crack sizes develops, with a significant probability that crack growth may reach dangerous sizes. This contradicts the deterministic concept, which expects that when $\ell_b \simeq \ell_h$ crack size growth is ~ 0. The probabilistic kinetics law represents the real, physical, probabilistic behavior, and the corresponding probabilistic equation gives a better design model.

(2) Stress corrosion cracking is usually described and analyzed with deterministic constitutive laws. However, it was shown that these analyses are approximations and the probabilistic theory yields better insight into the actual physical process. This is a major advantage in understanding and controlling the complex behavior of SCC.

(3) The rate-controlling mechanism of corrosion fatigue is thermally activated and, consequently, a random process. The probabilistic theory provides improved understanding of this major mode of fracture failure.

(4) Advanced materials are fast replacing conventional ones, resulting in new problems. The reliable, rationally derived prediction and extrapolation of crack growth velocity is one of the most important of these. Application of probabilistic fracture constitutive laws shows that the actual crack velocity may be significantly higher than predicted by the deterministic, conventional analysis. The discrepancy is worthy of consideration in improving design safety and reliability, particularly in aircraft design, where new structural materials must be as light as possible, and at the same time highly reliable; furthermore, to assure satisfactory financial returns, aircraft must operate for a long time. In addition, aircraft are exposed to a wide range of atmospheric and operating conditions. Physically based rational design thus calls for application of probabilistic constitutive laws.

Besides demonstrating the application of probabilistic fracture constitutive laws for the solution of important engineering problems, this chapter gave some indication of the value of probabilistic fracture kinetics as a method of fracture control.

References

Chapter 1

1. D. Broek: *Elementary Engineering Fracture Mechanics*, Sijthoff and Noordhoff International Publishers (1978).
2. A. S. Tetelman and A. J. McEvily, Jr.: *Fracture of Structural Materials*, John Wiley and Sons (1967).
3. F. A. McClintock and A. S. Argon: *Mechanical Behavior of Materials*, Addison-Wesley Publishing Co. (1966).
4. G. E. Dieter: *Engineering Design*, McGraw-Hill Book Co. (1983).
5. R. W. Hertzberg: *Deformation and Fracture Mechanics of Engineering Materials*, 2nd edn., John Wiley and Sons (1983).
6. A. S. Krausz and H. Eyring: *Deformation Kinetics*, Wiley-Interscience (1975).
7. A. H. Cottrell: *An Introduction to Metallurgy*, Edward Arnold Publishers (1967).
8. A. S. Krausz and K. Krausz: Time dependent failure of ceramic materials in sustained and fatigue loading, *Proc. of the Fourth International Symposium on the Fracture Mechanics of Ceramics* (R. C. Bradt, A. G. Evans, D. P. H. Hasselman and F. F. Lange, eds.), Plenum Press (1986), pp. 333—340.
9. E. R. Fuller, Jr. and R. M. Thomson: Lattice theories of fracture, *Fracture Mechanics of Ceramics*, Vol. 4 (R. C. Bradt, D. P. H. Hasselman and F. F. Lange, eds.), Plenum Press (1978), pp. 507—548.
10. A. S. Krausz and K. Krausz: The elastic-plastic crack propagation in ceramic materials: theory and test, *Proc. of the International Symposium on Plastic Deformation of Ceramic Materials* (R. E. Tressler and R. C. Bradt, eds.), Plenum Publishing Co. (1984), pp. 655—668.
11. A. S. Griffith: The phenomena of rupture and flow in solids, *Phil. Trans. Roy. Soc.* **A221** (1980), pp. 163—198.
12. K. Krausz and A. S. Krausz: The kinetics of time dependent plastic flow and fracture in ceramic materials, *Proc. of the International Symposium on Plastic Deformation of Ceramic Materials* (R. E. Tressler and R. C. Bradt, eds.), Plenum Publishing Co. (1984), pp. 547—554.
13. A. S. Krausz and K. Krausz: A review of the rate dependent plastic deformation processes and the associated constitutive laws, *Proc. of the International Conference on Constitutive Laws for Engineering Materials: Theory and Applications* (C. S. Desai and R. H. Gallagher, eds.), College of Engineering (1983), pp. 81—89.
14. M. F. Kanninen and C. H. Popelar: *Advanced Fracture Mechanics*, Oxford Press (1985).
15. H. Eyring, D. Henderson, B. J. Stover and E. M. Eyring: *Statistical Mechanics and Dynamics*, Wiley (1964).

16. S. Glasstone, K. J. Laidler and H. Eyring: *The Theory of Rate Processes*, McGraw-Hill (1941).
17. K. J. Laidler: *Chemical Kinetics*, McGraw-Hill (1965).
18. W. Jost: *Diffusion in Solids, Liquids, Gases*, Academic Press (1960).
19. P. G. Shewmon: *Diffusion in Solids*, McGraw-Hill (1963).
20. J. R. Manning: *Diffusion Kinetics for Atoms in Crystals*, Van Nostrand (1968).
21. J. Crank: *The Mathematics of Diffusion*, Oxford University Press (1956).
22. T. L. Hill: *An Introduction to Statistical Thermodynamics*, Addison-Wesley (1962).
23. A. S. Krausz: The theory of non-steady state fracture propagation rate, *Int. J. Fract.* **12** (1976), pp. 239—242.
24. J. W. Obreimoff: The splitting strength of mica, *Proc. Royal Soc. (London)* **A127** (1930), pp. 290—297.
25. E. Orowan: Die Zugfestigkeit von Glimmer und das Problem der Technischen Festigkeit, *Z. Physik* **82** (1933), pp. 235—255.
26. A. J. Forty and C. T. Forwood: The healing of cleavage cracks in alkali halide crystals, *Trans. Brit. Ceram. Soc.* **62** (1963), pp. 715—724.
27. S. M. Wiederhorn and P. R. Townsend: Crack healing in glass, *J. Am. Ceram. Soc.* **534** (1970), pp. 486—489.
28. B. R. Lawn and T. R. Wilshaw: *Fracture of Brittle Solids*, Cambridge University Press (1975).
29. A. S. Krausz and K. Krausz: Time and temperature dependent plastic flow and fracture: physical theory and engineering applications, *Proceedings of the XIIth Southeastern Conference on Theoretical and Applied Mechanics*, School of Engineering, Auburn University, Auburn, Al. (1984), pp. 169—179.
30. J.-C. Pollet and S. J. Burns: An analysis of slow crack propagation data in PMAA and brittle materials, *Int. J. Fract.* **13** (1977), pp. 775—789.

Chapter 2, Part 1

1. A. S. Krausz and H. Eyring: *Deformation Kinetics*, Wiley-Interscience (1975).
2. A. S. Krausz and H. Eyring: Chemical kinetics of plastic deformation, *J. Appl. Phys.* **42** (1971), p. 2382.
3. S. Glasstone, K. J. Laidler and H. Eyring: *The Theory of Rate Processes*, McGraw-Hill (1941).
4. S. W. Benson: *The Foundations of Chemical Kinetics*, McGraw-Hill (1960).
5. S. D. Brown: Multibarrier kinetics of brittle fracture: I, Stress dependence of the subcritical crack velocity, *J. Am. Ceram. Soc.* **62** (1979), pp. 515—524.
6. A. S. Krausz, K. Krausz and D. Necsulescu: Reliability theory of stochastic fracture processes in sustained loading: Part I, *Z. Naturforsch.* **38a** (1983), pp. 719—722.
7. A. S. Krausz and K. Krausz: The theory of thermally activated fracture: environment assisted crack propagation, *Canad. Metall. Quarterly* **23** (1984), pp. 107—113.
8. B. R. Lawn: An atomistic model of kinetic crack growth in brittle solids, *J. Mater. Sci.* **10** (1975), pp. 469—480.
9. R. Thomson: Theory of chemically assisted fracture, *J. Mater. Sci.* **15** (1980), pp. 1014—1034.
10. D. R. Clarke, B. R. Lawn and D. H. Roach: The role of surface forces in fracture, in *Fracture Mechanics of Ceramics*, Vol. 8 (R. C. Bradt, A. G. Evans and D. P. H. Hasselman, eds.), Plenum Press (1986), pp. 341—350.
11. S. D. Brown: A multibarrier kinetics approach to subcritical crack growth in glasses and ceramics, *Am. Ceram. Soc. Bull.* **55** (1976), pp. 395—000.
12. E. D. Case, J. R. Smyth and O. Hunter, Jr.: Microcrack healing during the temperature cycling of single phase ceramics, in *Fracture Mechanics of Ceramics*, Vol. 5 (R. C. Bradt, A. G. Evans and D. P. H. Hasselman, eds.), Plenum Press (1983), pp. 507—530.
13. B. C. Bunker and T. A. Michalske: Effect of surface corrosion on glass fracture, in *Fracture Mechanics of Ceramics*, Vol. 8 (R. C. Bradt, A. G. Evans and D. P. H. Hasselman, eds.), Plenum Press (1986), pp. 391—412.

14. S. M. Wiederhorn: Mechanisms of subcritical crack growth in glass, in *Fracture Mechanics of Ceramics*, Vol. 4 (R. C. Bradt, D. P. H. Hasselman and F. F. Lange, eds.), Plenum Press (1978), pp. 549—580.
15. A. Kelly: *Strong Solids*, Clarendon Press, Oxford (1966).
16. J. W. Obreimoff: The splitting strength of Mica, *Proc. Royal Soc.* **A127** (1930), pp. 290—297.
17. E. Orowan: Die Zugfestigkeit von Glimmer und das Problem der Technischen Festigkeit, *Z. Physik* **81** (1933), pp. 235—255.
18. A. J. Forty and C. T. Forwood: The healing of cleavage cracks in alkali halide crystals, *Trans. Brit. Ceram. Soc.* **62** (1963), pp. 715—724.
19. S. M. Wiederhorn and P. R. Townsend: Crack healing in glass, *J. Am. Ceram. Soc.* **534** (1970), pp. 486—489.
20. B. R. Lawn and T. R. Wilshaw: *Fracture of Brittle Solids*, Cambridge University Press (1975).
21. A. S. Krausz: The random walk theory of crack propagation, *Eng. Fract. Mech.* **12** (1979), pp. 499—504.
22. S. M. Wiederhorn, B. J. Hockey and D. E. Roberts: Effect of temperature on the fracture of sapphire, *Phil. Mag.* **28** (1973), pp. 783—796.
23. A. S. Krausz and K. Krausz: A fracture kinetics representation of fatigue crack propagation rate, *Proc. of the 2nd International Conference on Fatigue and Fatigue Thresholds* (C. H. Beevers, J. Bachlund, P. Lukas, J. Schijve and R. O. Ritchie, eds.) (1984), pp. 497—510.
24. T. O'D. Hanley and A. S. Krausz: Thermally activated deformation, I: Method of analysis, *J. Appl. Phys.* **45** (1974), p. 2013.
25. A. S. Krausz: Time-dependent crack propagation in linear-elastic solids, *J. Appl. Phys.* **49** (1978), pp. 3774—3778.
26. H. M. Cekirge, W. R. Tyson and A. S. Krausz: Static corrosion and static fatigue of glass, *J. Am. Ceram. Soc.* **58** (1976), pp. 265—266.
27. W. R. Tyson, H. M. Cekirge and A. S. Krausz: Thermally activated fracture of glass, *J. Mater. Sci.* **11** (1976), pp. 780—782.
28. R. Thomson, C. Hsieh and V. Rana: Lattice trapping of fracture cracks, *J. Appl. Phys.* **42** (1971), pp. 3154—3160.
29. S. D. Brown: The multibarrier kinetics of subcritical brittle crack growth, *Proc. of the 6th Canadian Congress on Applied Mechanics* (1977), pp. 257—258.
30. C. Hsieh and R. Thomson: Lattice theory of fracture and crack creep, *J. Appl. Phys.* **44** (1973), pp. 2051—2063.
31. B. R. Lawn: An atomistic model of kinetic crack growth in brittle solids, *J. Mater. Sci.* **10** (1973), pp. 469—480.
32. J.-C. Pollet and S. J. Burns: Thermally activated crack propagation theory, *Int. J. Fract.* **13** (1977), pp. 667—697.
33. J.-C. Pollet and S. J. Burns: An analysis of slow crack propagation data in PMAA and brittle materials, *Int. J. Fract.* **13** (1977), pp. 775—789.
34. S. M. Wiederhorn: Influence of water vapor on crack propagation in soda-lime glass, *J. Am. Ceram. Soc.* **50** (1967), pp. 407—414.
35. S. M. Wiederhorn: Subcritical crack growth in ceramics, in *Fracture Mechanics of Ceramics*, Vol. 2 (R. C. Bradt, D. P. H. Hasselman and F. F. Lange, eds.), Plenum Press (1974), pp. 613—646.
36. K. J. Laidler: *Chemical Kinetics*, McGraw-Hill (1965).
37. A. S. Krausz: The deformation and fracture kinetics of stress corrosion cracking, *Int. J. Fract.* **14** (1978), pp. 5—15.
38. A. S. Krausz: The theory of thermally activated processes in brittle stress corrosion cracking, *J. Eng. Fract. Mech.* **11** (1979), pp. 33—42.
39. S. H. Knickerbocker, A. Zangvil and S. D. Brown: Displacement rate and temperature effects in fracture of a hot-pressed silicon nitride at 1100 to 1325 °C, *J. Am. Ceram. Soc.* **67** (1984), pp. 365—368.
40. S. H. Knickerbocker, A. Zangvil and S. D. Brown: High-temperature mechanical properties and microstructure for hot-pressed silicon nitrides with amorphous and crystalline intergranular phases, *J. Am. Ceram. Soc.* **68** (1985), pp. C99—C101.

Chapter 2, Part 2

1. R. W. Hertzberg: *Deformation and Fracture Mechanics of Engineering Materials*, 2nd edn., John Wiley and Sons (1983).
2. R. Thomson: Theory of chemically assisted fracture (Part I) and E. R. Fuller and R. Thomson (Part 2), *J. Mater. Sci.* (1980), pp. 1014—1034.
3. H. E. Boyer and T. L. Gall (eds.): *Metals Handbook*, Desk edn., American Society for Metals (1985).
4. T. Yokobori: *An Interdisciplinary Approach to Fracture and Strength of Solids*, Wolters—Noordhoff Publishing (1968).
5. H. Liebowitz (ed.): *Fracture, and Advanced Treatise*, Vol. 3, *Engineering Fundamentals and Environmental Effects*, Academic Press (1971).
6. K. J. Laidler: *Chemical Kinetics*, McGraw-Hill (1965).
7. S. M. Wiederhorn and L. H. Boltz: Stress corrosion and static fatigue of glass, *J. Am. Ceram. Soc.* **53**(10) (1970), pp. 543—548.
8. B. R. Lawn and T. R. Wilshaw: *Fracture of Brittle Solids*, Cambridge University Press (1975).
9. R. J. Charles and W. B. Hillig: The kinetics of glass failure by stress corrosion cracking, *Mechanical Strength of Glass and Ways of Improving It*, Union Scientifique Continentale du Verre (1962), pp. 511—527.
10. W. B. Hillig and R. J. Charles: Surfaces, stress-dependent surface reactions and strength, *High-strength Materials* (F. Zackay, ed.), Wiley-Interscience (1964), pp. 682—701.
11. J. C. Scully (ed.): *The Theory of Stress Corrosion Cracking in Alloys*, NATO Scientific Affairs Division (1971).
12. S. M. Wiederhorn: Influence of water vapor on crack propagation in soda-lime glass, *J. Am. Ceram. Soc.* **50**(8) (1967), pp. 407—414.
13. J. K. Tien, A. W. Thompson, I. M. Bernstein and R. J. Richards: Hydrogen transport by dislocations, *Met. Trans. A.* **7A** (1976), pp. 821—829.
14. S. D. Brown: Multibarrier kinetics of brittle fracture: I, Stress dependence of subcritical crack velocity, *J. Am. Ceram. Soc.* **62** (1979), pp. 515—524.
15. J.-C. Pollet and S. J. Burns: Thermally activated crack propagation — theory, *Int. J. Fract. Mech.* **13** (1977), pp. 667—679.
16. S. D. Brown: A multibarrier rate process approach to subcritical crack growth, in *Fracture Mechanics of Ceramics* (R. C. Bradt, D. P. H. Hasselman and F. F. Lange, eds.), Vol. 4, Plenum Press (1978), pp. 597—621.
17. G. W. Powell and S. E. Mahmoud: Failure analysis and prevention, in *Metals Handbook* (9th edn.), Vol. 11, American Society for Metals (1986).
18. J. R. Newby (coordinator): *Metals Handbook* (9th edn.), Vol. 8, *Mechanical Testing*, American Society for Metals (1985).
19. A. S. Krausz: The theory of thermally activated processes in brittle stress corrosion cracking, *J. Eng. Fract. Mech.* **11** (1979), pp. 33—42.
20. H. H. Johnson: Hydrogen brittleness in hydrogen-oxygen gas mixtures, in *Stress Corrosion Cracking and Hydrogen Embrittlement of Iron Base Alloys*, NACE-5, National Association of Corrosion Engineers (1977), pp. 382—389.
21. S. J. Hudak and R. P. Wei: Hydrogen enhanced crack growth in 18 Ni maraging steels, *Metall. Trans.* **7A** (1976), pp. 235—241.
22. S. M. Wiederhorn, B. J. Hockey and D. E. Roberts: Effect of temperature on the fracture of sapphire, *Phil. Mag.* **28** (1973), pp. 783—796.
23. S. M. Wiederhorn: A chemical interpretation of static fatigue, *J. Am. Ceram. Soc.* **55**(2) (1972), pp. 81—85.
24. W. R. Tyson, H. M. Cekirge and A. S. Krausz: Thermally activated fracture of glass, *J. Mater. Sci.* **11** (1976), p. 780.
25. H. M. Cekirge, W. R. Tyson and A. S. Krausz: Static corrosion and static fatigue of glass, *J. Am. Ceram. Soc.* **58** (1976), p. 265.
26. A. S. Krausz: The deformation and fracture kinetics of stress corrosion cracking, *Int. J. Fract.* **14** (1978), pp. 5—15.

27. A. S. Krausz and K. Krausz: The inherently probabilistic character of subcritical fracture processes, *Trans. ASME, J. Mech. Design, Special Issue: Reliability, Stress Analysis, and Failure Prevention* **104** (1982), pp. 666—670.

28. A. S. Krausz and K. Krausz: A unified fracture kinetics representation of the three regions of stress corrosion cracking, *Int. J. Fract.* **23** (1900), pp. 169—175.

29. S. M. Wiederhorn, H. Johnson, A. M. Diness and H. A. Heuer: Fracture of glass in vacuum, *J. Am. Ceram. Soc.* **57**(8) (1974), pp. 336—341.

30. A. S. Krausz: Time dependent crack propagation in linear-elastic solids, *J. Appl. Phys.* **49** (1978), pp. 3774—3778.

31. S. W. Freiman: Effects of alcohols on crack propagation in glass, *J. Am. Ceram. Soc.* **57**(8) (1974), pp. 350—353.

32. A. G. Evans and M. Linzer: Failure prediction in structural ceramics using acoustic emission, *J. Am. Ceram. Soc.* **56** (1973), pp. 575—581.

33. S. M. Wiederhorn: Mechanisms of subcritical crack growth in glass, in *Fracture Mechanics of Ceramics*, Vol. 4 (R. C. Bradt, D. P. H. Hasselman and F. F. Lange, eds.), Plenum Press (1978), pp. 549—580.

34. K. Sadananda and P. Shahinian: Creep-fatigue crack growth, in *Cavities and Cracks in Creep and Fatigue* (J. Gittus, ed.), Applied Science Publishers (1981), pp. 109—195.

35. K. Sadananda and P. Shahinian: Evaluation of J* parameter for creep crack growth in type 316 stainless steel, in *Fracture Mechanics*, ASTM STP 791, Vol. 2 (J. C. Lewis and G. Sines, eds.), American Society for Testing and Materials (1983), pp. II-183—II-196.

36. F. Garofalo: *Fundamentals of Creep and Creep Rupture in Metals*, The Macmillan Co. (1965).

37. D. Broek: *Elementary Engineering Fracture Mechanics*, Sijthoff and Noordhoff International Publishers (1978).

38. B. Wilshire and R. J. Owen: *Engineering Approaches to High Temperature Design*, Pineridge Press (1983).

39. V. S. Kuksenko and V. P. Tamuzs: *Fracture Micromechanisms of Polymer Materials*, Martinus Nijhoff Publishers (1981).

40. J. Gittus: *Cavities and Cracks in Creep and Fatigue*, Applied Science Publishers (1981).

41. K. Sadananda, B. B. Rath and D. J. Michel: *Micro and Macro Mechanics of Crack Growth*, The Metallurgical Society of AIME (1982).

42. R. P. Wei and J. D. Landes: Correlation between sustained-load and fatigue crack growth in high-strength steels, *Mat. Res. Standards* **9**(7) (1969), pp. 25—27 and 44—46.

43. A. J. McEvily and R. P. Wei: Chemistry, mechanics and microstructure, in *Corrosion Fatigue*, NACE-2, National Association of Corrosion Engineers (1972), p. 381.

44. R. P. Wei and G. W. Simmons: Environment enhanced fatigue crack growth in high-strength steels, in *Stress Corrosion Cracking and Hydrogen Embrittlement of Iron Base Alloys*, NACE-5, National Association of Corrosion Engineers (1977), pp. 751—765.

45. R. P. Wei and G. Shim: Fracture mechanics and corrosion fatigue, in *Corrosion Fatigue: Mechanics, Metallurgy, Electrochemistry, and Engineering* (T. W. Crooker and B. N. Leis, eds.), ASTM STP 801, American Society for Testing and Materials (1983), pp. 5—25.

46. A. S. Krausz and K. Krausz: A fracture kinetics representation of fatigue crack propagation rate, in *Proc. of the 2nd International Conference on Fatigue and Fatigue Thresholds* (C. H. Beevers, J. Bachlund, P. Lukas, J. Schijve, and R. O. Ritchie, eds.) (1984), pp. 497—510.

47. N. E. Frost, K. J. Marsh and L. P. Pook: *Metal Fatigue*, Clarendon Press (1974).

48. S. T. Rolfe and J. M. Barsom: *Fracture and Fatigue Control in Structures*, Prentice-Hall (1977).

49. M. Klesnil and P. Lukas: *Fatigue of Metallic Materials*, Elsevier Scientific Publishing Co. (1980).

50. H. O. Fuchs and R. I. Stephens: *Metal Fatigue in Engineering*, John Wiley and Sons (1980).

51. T. V. Duggan and J. Byrne: *Fatigue as a Design Criterion*, The Macmillan Press Ltd. (1979).

52. C. C. Osgood: *Fatigue Design*, Pergamon Press (1982).

53. R. P. Skelton (ed.): *Fatigue at High Temperature*, Applied Science Publishers (1983).

54. R. D. Carter, E. W. Lee, E. A. Starke, Jr. and C. J. Beevers: The effect of microstructure and

environment on fatigue crack closure of 7475 aluminum alloy, *Metal. Trans.* **15A** (1984), pp. 555—563.

55. M. O. Speidel: Corrosion Fatigue in Fe-Ni-Cr Alloys, in *Stress Corrosion Cracking and Hydrogen Embrittlement of Iron Base Alloys*, NACE-5, National Association of Corrosion Engineers (1977), pp. 1071—1094.

56. O. Vosikovsky: Fatigue-crack growth in an X-65 line-pipe steel at low frequencies in aqueous environments, *Closed Loop, The Magazine of Mechanical Testing*, MTS Systems Corp., Vol. 6, No. 1 (1976), pp. 3—12.

57. W. W. Gerberich and K. A. Peterson: Micro and macro mechanics aspects of time-dependent crack growth, in *Micro and Macro Mechanics of Crack Growth* (K. Sadananda, B. B. Rath and D. J. Michel, eds.), The Metallurgical Society of AIME (1982), pp. 1—17.

58. W. W. Gerberich and N. R. Moody: A review of fatigue fracture topology effects on threshold and growth mechanisms, in *Fatigue Mechanisms* (J. T. Fong, ed.), ASTM STP 675, American Society for Testing and Materials (1979), pp. 292—341.

59. Y. Hirose and K. Tanaka: Nucleation and growth of stress-corrosion cracks in notched plates of high strength steels, in *Mechanical Behavior of Materials* ICM3, 2 (K. J. Miller and R. T. Smith, eds.), Pergamon Press (1979), pp. 409—419.

60. J. E. Campbell, W. W. Gerberich and J. H. Underwood: *Application of Fracture Mechanics for Selection of Metallic Structural Materials*, American Society for Metals (1982).

61. J. A. Collins: *Failure of Materials in Mechanical Design*, John Wiley and Sons (1981).

62. V. J. Colangelo and F. A. Heiser: *Analysis of Metallurgical Failures*, John Wiley and Sons (1974).

63. R. J. Charles: Stress rupture evaluations of high temperature structural materials, in *Fracture Mechanics of Ceramics*, Vol. 4 (R. C. Bradt, D. P. H. Hasselman and F. F. Lange, eds.), Plenum Press (1978), pp. 623—638.

64. K. Krausz, A. S. Krausz: The kinetics of time dependent plastic flow and fracture in ceramic materials, in *Deformation of Ceramic Materials* (R. E. Tressler and R. C. Bradt, eds.), Plenum Press (1984), pp. 547—554.

65. B. R. Lawn: An atomistic model of kinetic crack growth in brittle solids, *J. Mater. Sci.* **10** (1975), pp. 469—480.

66. A. S. Krausz and J. Mshana: Steady-state fracture kinetics of crack front spreading, *Int. J. Fract.* **19** (1982), pp. 227—293.

67. B. Faucher and A. S. Krausz: Properties of consecutive energy barriers and the associated behavior in plastic flow, *J. Appl. Phys.* **49** (1978), pp. 3774—3778.

68. A. S. Krausz: The deformation kinetics of consecutive energy barriers, *J. Phys. Chem. Solids* **40** (1979), pp. 351—356.

69. A. S. Krausz and B. Faucher: The steady-state kinetics of double-kink spreading, *Scripta Met.* **13** (1979), pp. 91—94.

70. A. S. Krausz and B. Faucher: The reaction kinetics for a system of consecutive energy barriers in plastic flow, *Advances in Molecular Relaxation and Interaction Processes* **18** (1980), pp. 11—20.

71. J. P. Hirth and J. Lothe: *Theory of Dislocations*, McGraw-Hill (1968).

72. I.-H. Lin and J. P. Hirth: On brittle crack advance by double-kink nucleation, *J. Mater. Sci.* **17** (1982), pp. 447—460.

73. R. Thomson, C. Hsieh and V. Rana: Lattice trapping of fracture cracks, *J. Appl. Phys.* **42** (1971), pp. 3154—3160.

74. C. Hsieh and R. Thomson: Lattice theory of fracture and crack creep, *J. Appl. Phys.* **44** (1973), pp. 2051—2063.

75. K. Krausz, A. S. Krausz and S.-S. Necsulescu: The effects of the internal stress field and the structure of crystalline materials on crack propagation velocity, *Int. J. Fract.* **23** (1983), pp. 155—159.

76. K. Krausz and A. S. Krausz: Stress corrosion life expectancy topographs for the determination of design and inspection data, in *Time-dependent Fracture, Proc. of the 11th Canad. Fracture Conf.* (A. S. Krausz, ed.), Martinus Nijhoff Publishers (1985), pp. 131—145.

77. S. N. Zhurkov: Kinetic concept of the strength of solids, *Int. J. Fract. Mech.* **1** (1965), pp. 311—323.

78. C. C. Hsiao: Fracture, *Phys. Today* **19** (1966), pp. 49—53.
79. F. P. Ford: Current understanding of the mechanism of stress corrosion and corrosion fatigue, in *Environment Sensitive Fracture* (S. W. Dean, E. N. Pugh and G. M. Vgiansky, eds.), ASTM STP 821, American Society for Testing and Materials (1984), pp. 32—51.
80. E. R. Fuller and R. M. Thomson: Lattice theories of fracture, in *Fracture Mechanics of Ceramics*, Vol. 4 (R. C. Bradt, D. P. H. Hasselman and F. F. Lange, eds.), Plenum (1983), pp. 507—548.
81. M. F. Kanninen and C. H. Popelar: *Advanced Fracture Mechanics*, Oxford Press (1985).
82. R. W. Hertzberg and J. A. Manson: *Fatigue of Engineering Plastics*, Academic Press (1980).
83. A. S. Argon: Thermally activated crack growth in brittle solids, *Scripta Met.* **16** (1982), pp. 259—264.
84. A. S. Krausz and K. Krausz: Time dependent failure of ceramic materials in sustained and fatigue loading, *Fracture Mechanics of Ceramics* (R. C. Bratt, A. G. Evans, D. P. H. Hasselman and F. F. Lange, eds.) **8** (1986), pp. 330—340.
85. K. Krausz and A. S. Krausz: Fracture kinetics of corrosion fatigue, *Third International Conference on Fatigue and Fatigue Thresholds*, University of Virginia, Charlottesville, VA, U.S.A. June 28—July 3 (1987).
86. A. S. Krausz and K. Krausz: The fracture kinetics of subcritical environment assisted fatigue crack propagation processes, ASTM *20th National Symposium on Fracture Mechanics: Perspectives and Directions*, Lehigh University, Bethlehem, PA, U.S.A. June 23—25 (1987).
87. T. Ungsuwarungsri and W. G. Knauss: The role of damage-softened material behaviour in the fracture of composites and adhesives, *Int. J. Fract.* **35** (1987), pp. 221—241.

Chapter 3, Part 1

1. R. A. Heller (ed.): *Probabilistic Aspects of Fatigue*, ASTM STP 511, American Society for Testing and Materials (1971).
2. R. E. Little and J. C. Ekvall (eds.): *Statistical Analysis of Fatigue Data*, ASTM STP 744, American Society for Testing and Materials (1979).
3. J. M. Bloom and J. C. Ekvall (eds.): *Probabilistic Fracture Mechanics and Fatigue Methods: Applications for Structural Design and Maintenance*, ASTM STP 798, American Society for Testing and Materials (1981).
4. W. Feller: *An Introduction to Probability Theory and its Applications*, Vol. 1, Wiley (1970).
5. D. T. Philips, A. Ravindron and J. J. Solberg: *Operations Research: Principles and Practice*, Wiley (1976).
6. A. S. Krausz and K. Krausz: The inherently probabilistic character of subcritical fracture processes, *Trans. Soc. Mech. Eng. Design, Special Issue: Reliability, Stress Analysis, and Failure Prevention* **104** (1982), pp. 666—670.
7. A. S. Krausz, K. Krausz and D. Necsulescu: Reliability theory of stochastic fracture processes in sustained loading: Part I, *Z. Naturforsch.* **38a** (1983), pp. 719—502.
8. Takeo Yokobori: Fracture, fatigue and yielding of materials as a stochastic process, *Kolloid Z.* **166**(1) (1959), pp. 20—24.
9. S. D. Brown: Multibarrier kinetics approach to subcritical crack growth in glasses and ceramics, *Am. Ceram. Soc. Bull.* **55**(4) (1976), pp. 395—401.
10. A. S. Krausz and K. Krausz: The theory of non-steady state fracture kinetics analysis, Part 1: General theory of crack propagation, *Eng. Fract. Mech.* **13** (1980), pp. 751—758.
11. A. S. Krausz, J. Mshana and K. Krausz: The theory of non-steady state fracture kinetics analysis; Part II: Non-steady state crack propagation in stress corrosion cracking, *Eng. Fract. Mech.* **13** (1980), pp. 759—766.
12. A. S. Krausz: Crack-size distribution in homogeneous solids, *Int. J. Fract.* **15** (1979), pp. 337—342.
13. A. S. Krausz and J. Mshana: Steady-state fracture kinetics of crack front spreading, *Int. J. Fract.* **19** (1982), pp. 277—293.

Chapter 3, Part 2

1. W. Feller: *An Introduction to Probability Theory and its Applications*, Vol. 1, Wiley (1970).
2. D. T. Philips, A. Ravindran and J. J. Solberg, *Operations Research: Principles and Practice*, Wiley (1976).
3. A. B. Clarke and R. L. Disney: *Probability and Random Processes for Engineers and Scientists*, Wiley (1970).
4. N. Wax (ed.): *Selected Papers on Noise and Stochastic Processes*, Dover (1954).
5. A. S. Krausz: Crack-size Distribution in Homogeneous Solids, *Int. J. Fract.* **15** (1979), pp. 337—342.
6. W. Jost: *Diffusion in Solids, Liquids, Gases*, Academic Press (1960).
7. P. G. Shewmon: *Diffusion in Solids*, McGraw-Hill (1963).
8. J. Crank: *The Mathematics of Diffusion*, Oxford University Press (1956).
9. P. C. Jordan: *Chemical Kinetics and Transport*, Plenum Press (1979).
10. R. M. Barrer: *Diffusion in and Through Solids*, Cambridge (1941).
11. J. R. Manning: *Diffusion Kinetics for Atoms in Crystals*, D. Van Nostrand (1968).
12. A. S. Krausz, K. Krausz and D. Necsulescu: Reliability theory of stochastic fracture processes in sustained loading: Part I, *Z. Naturforsch.* **38a** (1983), pp. 719—722.
13. A. S. Krausz: The random walk theory of crack propagation, *Eng. Fract. Mech.* **12** (1979), pp. 499—504.
14. W. J. Moore, *Physical Chemistry*, 3rd edn., Prentice-Hall (1962).
15. A. S. Krausz and K. Krausz: The theory of non-steady state fracture kinetics analysis, Part 1: General theory of crack propagation, *Eng. Fract. Mech.* **13** (1980), pp. 751—758.
16. A. S. Krausz, J. Mshana and K. Krausz: The theory of non-steady state fracture kinetics analysis; Part II: Non-steady state crack propagation in stress corrosion cracking, *Eng. Fract. Mech.* **13** (1980), pp. 759—766.
17. A. S. Krausz and K. Krausz: A unified fracture kinetics representation of the three regions of stress corrosion cracking, *Int. J. Fract.* **23** (1983), pp. 169—175.
18. A. S. Krausz and K. Krausz: A fracture kinetics representation of fatigue crack propagation rate, *Proc. of the 2nd International Conference on Fatigue and Fatigue Thresholds* (C. H. Beevers, J. Bachlund, P. Lukas, J. Schijive and R. O. Ritchie, eds.) (1984), pp. 497—510.
19. R. P. Wei and J. D. Landes: Correlation between sustained-load and fatigue crack growth in high-stress steels, *Mat. Res. Standards* **9**(7) (1969), pp. 25—27.
20. A. J. McEvily and R. P. Wei: in *Corrosion Fatigue: Chemistry, Mechanics and Microstructure*, NACE-2, National Association of Corrosion Engineers (1973), p. 381.
21. R. P. Wei and G. W. Simmons: Environment enhanced fatigue crack growth in high-strength steels, *Stress Corrosion Cracking and Hydrogen Embrittlement of Iron Base Alloys* (R. W. Staehle, J. Hochman, R. D. McCright and J. E. Slater, eds.), NACE Vol. 5, National Association of Corrosion Engineers (1977), pp. 751—765.
22. R. P. Wei and G. Shim: *Fracture Mechanics and Corrosion Fatigue*, (T. W. Crooker and B. N. Leis, eds.), ASTM STP 801, American Society for Testing and Materials (1983), pp. 5—25.
23. M. O. Speidel: Corrosion Fatigue in Fe-Ni-Cr Alloys, *Stress Corrosion Cracking and Hydrogen Embrittlement in Iron Base Alloys* (R. W. Staehle, J. Hochman, R. D. McCright and J. E. Slater, eds.), NACE Vol. 5, National Association of Corrosion Engineers (1977), pp. 1071—1094.
24. B. R. Lawn and T. R. Wilshaw: *Fracture of Brittle Solids*, Cambridge University Press (1975).
25. A. S. Krausz and K. Krausz: The theory of thermally activated fracture: environment assisted crack propagation, *Canad. Metall. Quarterly* **23** (1984), pp. 107—113.
26. A. G. Evans: A method for evaluating the time-dependent failure characteristics of brittle materials — and its application to polycrystalline alumina, *J. Mater. Sci.* **7**(10) (1972), pp. 1137—1146.
27. R. D. Carter, E. W. Lee, E. A. Starke, Jr. and C. J. Beevers: The effect of microstructure and environment on fatigue crack closure of 7475 aluminum alloy, *Metall. Trans.* **15A** (1984), pp. 555—563.

28. W. W. Stinchcomb and K. L. Reitsnider: Fatigue damage mechanisms in composite materials: A review, in *Fatigue Mechanisms* (J. T. Fong, ed.), ASTM STP 675, American Society for Testing and Materials (1979), pp. 763—787.
29. A. S. Krausz, K. Krausz ad D.-S. Necsulescu: *Probabilistic Fracture Kinetics of 'Natural' Composites*, (H. T. Hahn, ed.), ASTM STP 907, American Society for Testing and Materials (1986), pp. 73—83.
30. M. F. Kanninen and C. H. Popelar: *Advanced Fracture Mechanics*, Oxford University Press (1985).
31. S. W. Tsai and H. T. Hahn: *Introduction to Composite Materials*, Technomic (1980).
32. A. Kelly: *Strong Solids*, Oxford University Press (1973).
33. M. R. Pigott: *Load Bearing Fibre Composites*, Pergamon Press (1980).
34. A. S. Argon: Fracture of composites, in *Treatise on Materials Science and Technology*, Vol. 1 (H. Herman, ed.), Academic Press (1972), pp. 79—114.
35. V. A. Petrov and A. N. Orlov: Statistical kinetics of thermally activated fracture, *Int. J. Fract.* **12** (1976), pp. 231—238.
36. H. Ishikawa, A. Tsurui and A. Utsumi: A stochastic model of fatigue crack growth in consideration of random propagation resistance, *Proc. of the 2nd International Conference on Fatigue and Fatigue Thresholds* (C. H. Beevers, J. Bachlund, P. Lukas, J. Schijve and R. O. Ritchie, eds.) (1984), pp. 511—520.
37. R. Arone: A stochastic model for fatigue crack growth, *Proc. of the 2nd International Conference on Fatigue and Fatigue Thresholds* (C. H. Beevers, J. Bachlund, P. Lukas, J. Schijve and R. O. Ritchie, eds.) (1984), pp. 521—527.
38. S. B. Batdorf and H. L. Heinisch, Jr.: Fracture statistics of brittle materials with surface cracks, *Eng. Fract. Mech.* **11** (1978), pp. 831—891.
39. S. Aoki and M. Sakata: Statistical approach to delayed failure of brittle materials, *Int. J. Fract.* **16** (1980), pp. 459—469.

APPENDICES

APPENDICES

Review of fracture mechanics

The Griffith theory is the source of modern fracture theory: it has established that crack growth originates at defects which are inherent in all real materials. This growth occurs when the energy needed for the creation of new surfaces (and a plastic zone) is exceeded by the energy that is available from the external work performed during crack growth and from the change in the stored elastic energy. When this condition is satisfied, the crack grows catastrophically, at a significant fraction of the velocity of sound.

These concepts are now part of the fundamental conditions which define the model of fracture mechanics, although in forms that are modified to correct some of the shortcomings of the theory: it is established that a definite condition must exist in the crack-tip zone; the linear elastic model is extended to non-linear and to plastic solid continuum; and a mathematical formulation is developed that provides the rigorous basis for the practical measurements of the characteristic values which represent the overall, general characteristic fracture resistance of a specific material.

This description of the model and purposes of fracture mechanics is formulated to define its role in fracture kinetics. The following review presents a brief outline of fracture mechanics to facilitate the derivation and use of the fracture constitutive equations. It does not serve as an introduction to fracture mechanics; for this, the literature has to be consulted [1—7].

Linear Elastic Fracture Mechanics (LEFM)

The conditions that control the fracture of linear elastic solid continuum was developed first: understandably, because this is the simplest in behavior and mathematically the most accessible model. LEFM is now an established design and test engineering method. Its concepts and techniques are imprinted even on models that are developed for the fracture analysis of elastic—plastic, plastic and viscoplastic solids.

LEFM considers that while the energy criterion has to be satisfied, cracks can propagate only when a specific, critical stress condition exists in the

149

immediate neighborhood of the crack tip. Its aim is to define rigorous fracture design and test conditions according to the usual engineering practices. For example, it is a usual engineering practice to define the onset of plastic flow in any small volume element of a component subjected to a complex stress—strain field by a mathematically rigorous model — usually the Tresca (a stress) or the Mises (an energy) condition: wherever the characteristic value defined by the model is the same in both the complex stress condition found in the design practice and in the simple test conditions, then the material is in the same state. That is, plastic flow starts in a small volume of a component, subject to a complex stress field, if the characteristic value (Tresca or Mises) in this volume is equal to the characteristic value (calculated for the same model) measured in a simple test (tensile for instance) at yield. With this principle engineers can design simple tests to measure material properties for specific complex applications. LEFM provides the theory to follow the same practice of using simple tests to define the fracture state in complex conditions.

In a homogeneous linear elastic isotropic solid continuum containing a sharp crack (Figure A.1) the two-dimensional stress field is described as

$$\sigma_x = \frac{K}{(2\pi r)^{1/2}}\, f_1(\theta),$$

$$\sigma_y = \frac{K}{(2\pi r)^{1/2}}\, f_2(\theta), \tag{A.1}$$

$$\tau_{xy} = \frac{K}{(2\pi r)^{1/2}}\, f_3(\theta),$$

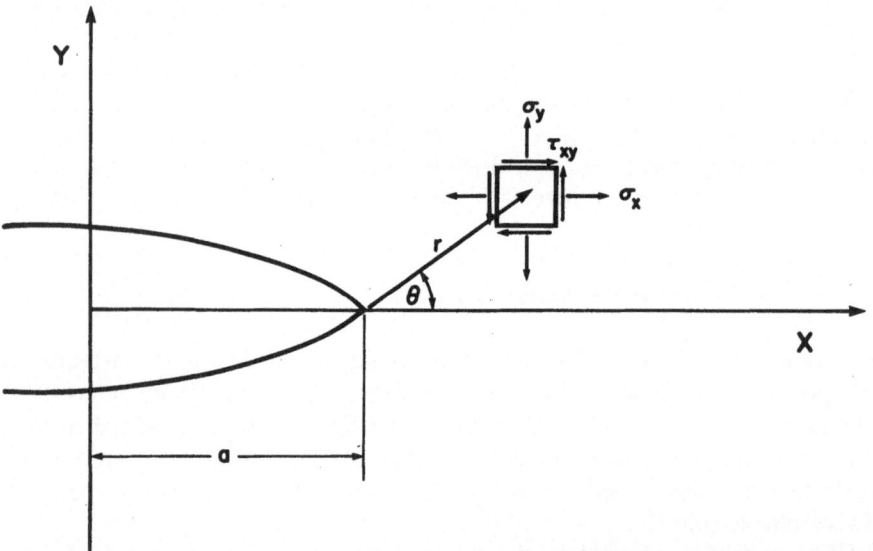

Figure A.1.   The coordinate system and stresses associated with the description of LEFM (equations (A.1)).

where σ_x, σ_y, τ_{xy} are the normal and shear stress components, r and θ are the polar coordinates. Equations (A.1) demonstrate that the effects of the load and the geometrical boundary conditions depend only on — and are fully expressed by — the stress intensity factor:

$$K = Y\sigma(\pi a)^{1/2}.$$

Here σ is the stress in the region where the effect of the crack is negligible. The factor Y is the function of the load distribution and the geometry of the component and the crack; for instance, in an infinite plate with a through crack and loaded in one-dimensional tension the stress intensity factor is

$$K = \sigma(\pi a)^{1/2}. \tag{A.2}$$

It follows from equations (A.1) that the stress distribution in two solids is the same if K is the same, although they may be loaded differently and be of different shape and size. According to fracture mechanics the crack propagates in a complexly shaped and loaded component if it propagates in a simple test at the same stress intensity factor K.

LEFM is now well established: it provides the rigorous description of the effects of macroscopic design quantities in the form

$$K = K \text{ (load, geometry)},$$

and it assures that equal values of K represent equal effects of these design quantities. This aspect of LEFM is essential, even fundamental, to fracture kinetics.

An alternative expression to K was also defined in LEFM. In the process of demonstrating that the LEFM model satisfies the Griffith energy condition the concept of elastic strain energy release rate per crack tip, also called crack extension force, was defined as

$$\mathscr{G} = \frac{1}{2}\frac{\mathrm{d}\Delta U_e}{\mathrm{d}a}.$$

The stress intensity factor K and the crack extension force $\mathscr{G}$ are interrelated: $\mathscr{G} = K^2/E$ for plane stress and $\mathscr{G} = (1 - \nu^2)(K^2/E)$, for plane strain, where E is Young's modulus and ν is Poisson's ratio.

It follows from equations (A.1) that close to the crack tip the stress increases to unrealistically high values; however, before this could occur, plastic yield takes place and the elastic model loses validity.

Plastic fracture mechanics

Figure A.2 shows that equation (A.1) loses validity at a distance r_y from the crack tip because the stress reaches the yield limit: inside this region plastic deformation occurs. While the yield limit is the function of the material, the temperature, and the deformation rate (and therefore, of the crack velocity), the extent of the plastic zone is a complex function of the environment as well.

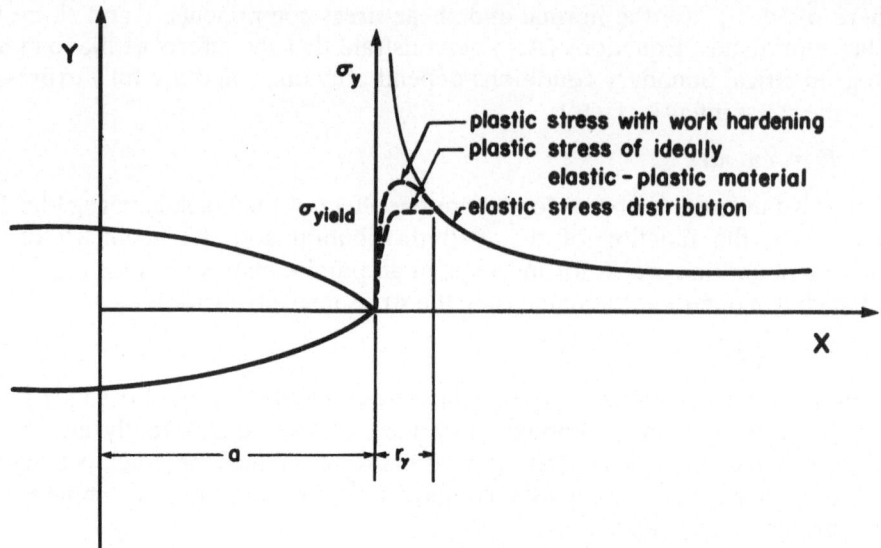

Figure A.2.   The plastic stress, plastic zone, and the uncorrected elastic stress distribution.

Yield starts where $\sigma_y = \sigma_{\text{yield}}$. From equation (A.1) it follows that along the x-axis

$$\sigma_{\text{yield}} = \sigma_y = \frac{K}{(2\pi r_y)^{1/2}}$$

and hence the plastic zone begins at the distance

$$r_y = \frac{1}{2\pi} \left(\frac{K}{\sigma_{\text{yield}}} \right)^2$$

from the tip. As Figure A.3 shows, this distribution would not result in an equilibrium of forces and, consequently, the elastic stress field becomes redistributed. The corrected plastic zone size, r_{cor}, was shown to be $r_{\text{cor}} = 2r_y$. In ductile materials there is always a plastic zone: it is the general practice to consider that when $a > 2.5\,(K/\sigma_{\text{yield}})^2$ linear elastic fracture mechanics can be used [8]. For plastic zones larger than this, the appropriate theories of fracture mechanics have to be applied.

Plastic deformation is a complex process, much more so than elastic deformation. The understanding of material behavior, and the mathematical techniques and their applications are less well developed. Consequently, crack propagation in elastic–plastic, and fully plastic conditions is defined also less well than in a linearly elastic solid. This is reflected by the concurrent use of several plastic fracture mechanics models; they are closely related, even derivatives, of LEFM. Three of the major ones — the J-integral, C^* and $\dot{J}$ — are of the principle and form that renders them very suitable for fracture kinetics and will be discussed from that point of view.

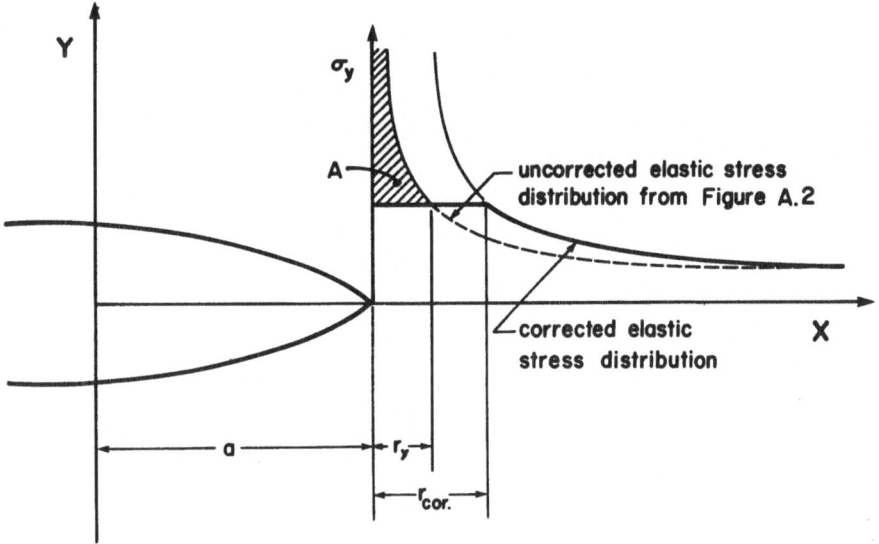

Figure A.3. The corrected stress distribution. The elastic stress field now balances the area *A*.

The J-integral method

The *J*-integral concept was originally developed as a rigorous energy condition of fracture analysis for non-linear elastic solids. Its extension for application to plastic deformation associated cracking is under active development but it has already reached the state where its use is a recommended industrial standard [9].

The *J*-integral is defined as (Figure A.4)

$$J = \int_S \left(W \, dy - \overline{T} \cdot \frac{\partial \bar{u}}{\partial x} \, ds \right)$$

where W is the loading work per unit volume (or for elastic bodies, strain energy density); S the path of the integral which encloses (that is, contains) the crack tip; ds is the increment of the contour path; $\overline{T}$ is the outward traction vector on ds; $\bar{u}$ is a displacement vector at ds; x,y are rectangular coordinates; and $(\partial \bar{u}/\partial x) \overline{T} \cdot ds$ is the rate of work input from the stress field into the area enclosed by S.

The *J*-integral is alternatively expressed by a more functional definition: "the *J*-integral is equal to the value obtained from two identical bodies with infinitesimally differing crack areas, each subject to stress, as the difference in loading work per unit difference in crack area at a fixed value of displacement, or where appropriate, at a fixed value of load. This approach is often used to define the

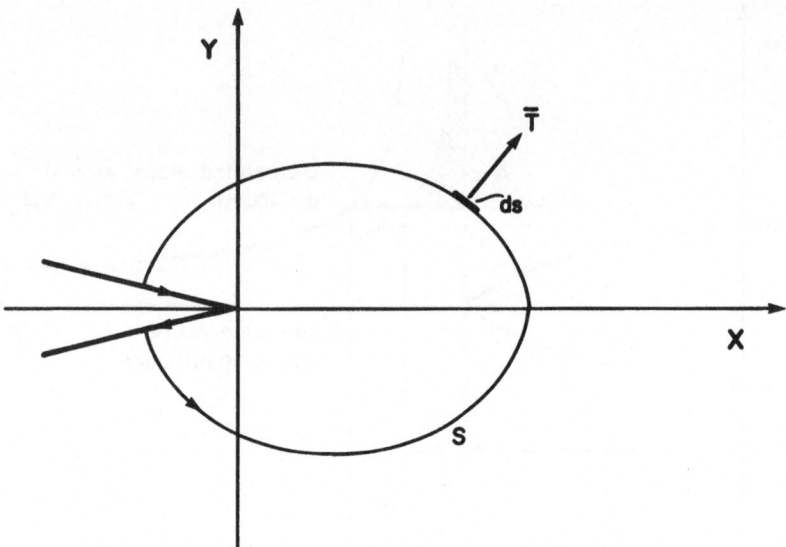

Figure A.4.   The line integral for the determination of the *J*-integral is defined as the figure shows.

value of *J* for inelastic solids" [9]. Accordingly,

$$J = -\frac{1}{B}\frac{\partial U}{\partial a},$$ (A.4)

where *B* is the specimen thickness. In equation (A.4), the work input, *U*, is determined from the measurement of the applied load at the load-point displacement. This form of the *J*-integral is eminently appropriate to the energy-based analysis of fracture kinetics.

It was noted that the *J*-integral was derived originally for a non-linear elastic model to determine the rate of potential energy change as a function of crack length, for a plate of unit thickness: it is then a concept similar to the crack extension force, $\mathcal{G}$, of the linear elasticity model. One of the consequences is that the development is valid only for elastic materials in which the loading— unloading paths coincide, and strictly speaking, it loses validity with increasing plastic zone size.

Similarly to the stress intensity factor concept, a specific value of *J* characterizes the fracture state: it is used in the same manner as *K*.

Time-dependent plastic flow

The *J*-integral is an appropriate representation of the factors that affect the crack propagation process for time-independent plastic deformation conditions. Because at high temperatures plastic flow is strongly time dependent this effect has to be considered also. For this an energy rate concept is defined

$$C^* = -\frac{1}{B}\frac{dU}{da},$$

where $d\dot{U}$ is the energy dissipation rate in creep during a crack extension of da, in a plate of thickness B. The testing, operational, description for uniaxial creep (when the stress, strain and time are separable) is

$$C^* = \frac{1}{B}\frac{P\dot{\delta}}{w}Y,$$

where w is the width of the specimen, P is the load, $\dot{\delta}$ is the load-point displacement rate, and Y is a non-dimensional factor

$$Y = Y(\text{specimen geometry, stress exponent } n)$$

for creep described as

$$\dot{\varepsilon} \propto \sigma^n.$$

References

1. A. S. Tetelman and A. J. McEvily: *Fracture of Structural Materials*, Wiley (1967).
2. J. F. Knott: *Fundamentals of Fracture Mechanics*, Wiley (1973).
3. S. T. Rolfe and J. M. Barsom: *Fracture and Fatigue Control in Structures; Applications of Fracture Mechanics*, Prentice-Hall (1977).
4. G. P. Cherepanov: *Mechanics of Brittle Fracture*, McGraw-Hill (1979).
5. R. W. Hertzberg: *Deformation and Fracture Mechanics of Engineering Materials* (2nd edn.), Wiley (1983).
6. K. Hellan: *Introduction to Fracture Mechanics*, McGraw-Hill (1984).
7. M. F. Kanninen and C. H. Popelar: *Advanced Fracture Mechanics*, Oxford University Press (1985).
8. *Annual Book of ASTM Standards*, E399—81, American Society for Testing and Materials, Part 10 (1982).
9. *Annual Book of ASTM standards*, E813—81, American Society for Testing and Materials, Section 3, Vol. 03.01 (1984).

Review of rate theory

The transition state theory established the description of the rate of atomic interaction changes. While several other theories were also developed this has the advantage of great formal simplicity; it is widely used in the description of fracture, plastic deformation, diffusion and chemical reactions. It is rigorously derived from the first principles of quantum statistical mechanics with assumptions that do not interfere with the validity of the description of the behavior of non-dynamic crack growth processes.

The following discussion is in a rather simplified form; for a full, rigorous treatment the references are offered.

A particularly suitable, picturesque, representation was given by Wigner and Eyring [2, 11]:

The motion of atoms can be conveniently represented in the 'configuration space'. Suppose that there are three atoms moving with respect to each other in a straight line. The configuration space is then two-dimensional, the X-coordinate being the distance of atoms 1 and 2, the Y-coordinate being the distance between 2 and 3. Every point in this configuration space corresponds to a configuration of the three atoms. The forces between these atoms can be derived from the potential energy for this configuration. If we make a landscape over the configuration space such that the height at any point is equal to the potential energy for this configuration, a ball rolling on this landscape will represent the motion of the three atoms under the influence of the forces among them (Figure B.1). The relative position of the atoms will change in real space in such a way that the corresponding point in configuration space always coincides with the position of the ball. If we are interested in a system of more than three atoms, or if their motion is not restricted to a line, we must use a configuration space of more than three dimensions.

Stable crack states correspond to low regions in our landscape. If the ball is in such a low region and has little velocity, it will stay in this region forever. There may be several regions of comparatively low energy, corresponding to several apparently stable groups of atoms. A reaction will then consist in the passing over of our ball from one low region to another.

For such a passage it needs, first of all, enough energy. The average amount

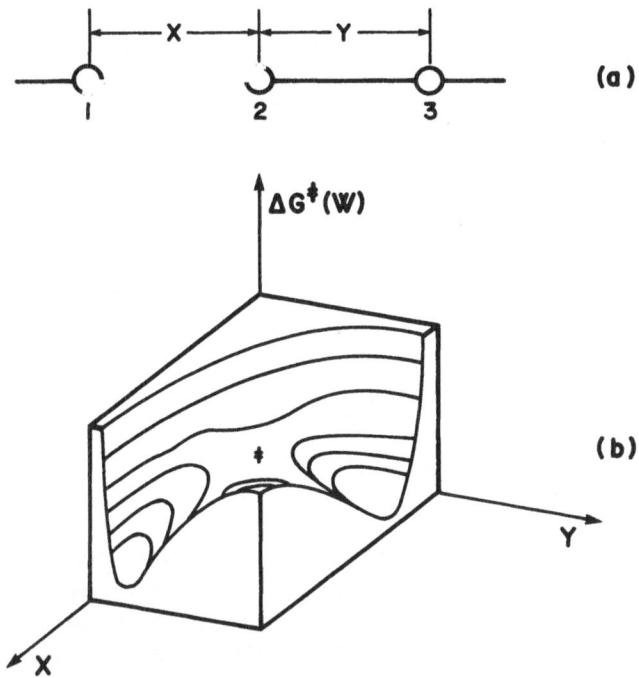

Figure B.1.   (a) The coordinates of three atoms moving in a straight line; (b) a typical potential energy surface for the linear three-atom interaction shown in (a).

of energy of such systems is proportional to the temperature. However, at any particular temperature some systems will have less than the average energy, and a very small number very much more energy. Only the systems with exceptionally high energy will be able to pass from one low region to the other, and the fraction of the systems that have this unusual amount of energy increases very rapidly with the temperature. This fact accounts for the rough empirical rule that the reaction rate doubles with an increase in the temperature by 10 degrees and it is one of the essential concepts of thermal activation.

The instantaneous state of each crack-tip zone can be represented by a point in the configuration space, with one point in the configuration space for each. As the crack tips move, the corresponding points in the configuration space will swarm like people in a mountainous region. To begin with, only one valley in configuration space is populated. The people scurry around apparently aimlessly. Most of them have too little energy to rise much above the floor of the valley. Even those who have enough energy to emerge from the crowd will go uphill at any arbitrary place, and only a few of the lucky ones will strike the path that leads into the neighboring valley. The number of these successful ones is all that concerns us.

We can count these lucky people by multiplying their number at the top of the pass by the velocity with which they are travelling. Account must be taken, of course, of the fact that some of those who pass the crest of the hill, having encountered some obstacle, return without descending into the new valley.

Both the number of the people in the pass and their velocity are given by standard formulas of statistical mechanics. Indeed, the velocity depends only on the temperature and the mass of the atoms involved. The number depends only on the temperature, the density of population in the valley, and the height of the pass above the valley floor [1, 3, 5, 7—9].

Of all the quantities entering into the calculation of reaction rates, the height of the pass above the valley is responsible for the greatest uncertainty. Except for a few cases, so far, it has always been necessary to derive this value from experiments, and only approximate values could be calculated theoretically. The question of absolute rates thus reduces to the construction of the appropriate energy surfaces, after which it becomes a relatively simple problem of arithmetic.

The absolute rate theory

Quantum mechanical calculation was used to construct potential energy surfaces for a wide variety of processes. With the development of these surfaces in configuration space, the stage was set for applying statistical mechanics in an unambiguous fashion to rate theory. The potential energy surface involves as many dimensions as are required to define a configuration plus one more for the energy. For a non-linear system of three atoms this means a four-dimensional space; in general, $3N-5$ dimensions are required for N atoms. For a practical material that is subjected to mechanical effect, the potential energy surface is roughly of 10^{22} dimensions. Besides the lowest surface on which most reactions take place, there may be an infinity of higher, excited surfaces.

On the potential surface the stable cracks are in the valleys. Two valleys are connected by a pass that leads through a saddle point. The saddle point is the place of almost no return and is called the transition or activated state. The atoms of the process zone in the configuration that corresponds to the transition or activated state is called the transition or activated complex. The path along which the systems travel from one valley to the other is the activation path and the distance along a valley and up through the pass into a neighboring valley is said to be the distance along the reaction coordinate. A few possible activation path types are shown in Figure B.2. In (a) the simple, one-barrier activation path is shown corresponding to the energy surface represented in Figure B.1. In

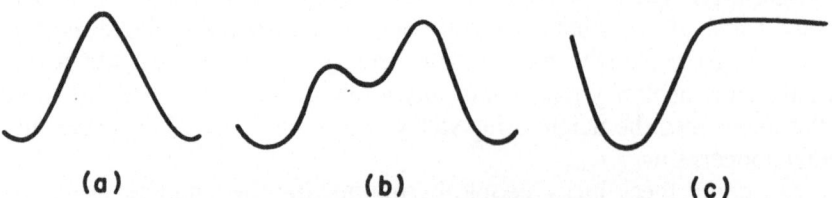

(a)   (b)   (c)

Figure B.2.   Activation paths of three different processes.

(b) a process is represented in which an intermediate stage is present; (c) represents one step in a fracture process. The atoms, under the effect of the stress, leave the stable configuration of the valley and move onto the flat plateau of the energy curve.

The system sitting in a low-lying region is in a stable configuration. If this region is separated from all other low places by regions higher than ~1 eV, then it is stable below room temperature. The higher the lowest pass, the higher is the temperature at which the crack is still stable. A reaction corresponds to a system moving from one low region to another. In thermally activated processes the Boltzmann factor makes it certain that the reaction will proceed by way of the lowest pass.

By definition, the activated state is always a saddle point with positive curvature in all degrees of freedom except the one that corresponds to crossing the barrier for which it is, of course, negative (Figure B.3).

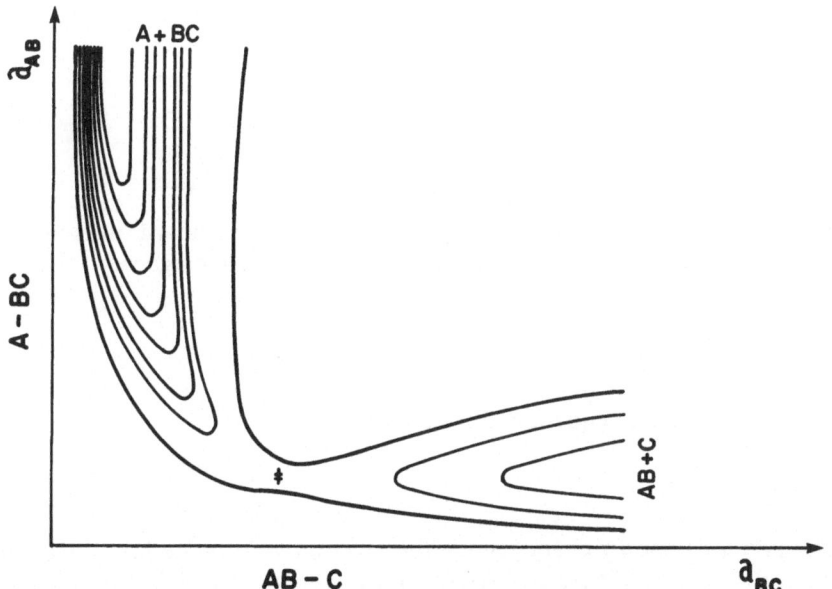

Figure B.3.   The potential energy surface of a linear three-atom system. This figure illustrates the shape of the energy surface at the saddle-point configuration.

By assuming an equilibrium to exist between the initial and the activated states, the specific rate of a reaction can be determined by calculating the concentration of the activated complexes and their rate of passage across the saddle point using the methods of statistical mechanics.

The rate of reaction over a single potential barrier and in the forward direction is

$$\text{rate}_f = C_f^{\ddagger} \frac{\bar{v}_f}{\Delta_f} \qquad (B.1)$$

where C_f^+ is the concentration of activated complexes per unit volume lying in the length Δ along the reaction path at the activated state. The activated state is reached when the complex is at the saddle point within the activation distance Δ. The average velocity of the activated complexes moving within the distance Δ is $\bar{v}$. Thus the ratio $\bar{v}/\Delta$ represents the frequency of emptying the length of path Δ of activated complexes.

The only condition on the length of the activation distance Δ is that it should be short enough to represent well the concept described by equation (B.1). Anticipating the development of equation (B.1), it is convenient to define the activation distance as

$$\Delta = h(2\pi m^+ \cdot kT)^{-1/2}$$

where h is Planck's constant, m^+ is the mass of the activated complex in motion through the distance Δ, k is the Boltzmann constant, and T is the absolute temperature. The mean velocity of the translational motion can be expressed as

$$\bar{v} = \frac{\displaystyle\int_0^\infty \exp\left(-\frac{1}{2} m^+ v^2/kT\right) v \, dv}{\displaystyle\int_{-\infty}^\infty \exp\left(-\frac{1}{2} m^+ v^2/kT\right) dv}, \tag{B.2}$$

when the energy levels are closely spaced. Using standard integral tables, equation (B.2) results in the following expression for the average velocity:

$$\bar{v} = \left(\frac{kT}{2\pi m^+}\right)^{1/2}.$$

Substituting Δ and $\bar{v}$ in equation (B.1) leads to

$$\text{rate}_f = \frac{kT}{h} C_f^+. \tag{B.3}$$

The potential energy change along the reaction coordinate for an elementary reaction is shown in Figure B.4. The zero of energy is taken as the ground state of the reactants, and ΔG_f^+ is the activation energy in the forward direction at absolute zero.

Assuming that the reactants are in equilibrium with the activated complexes, it can be shown that the same number of activated complexes are crossing the barrier in each direction. It is important to realize that there is no interaction between activated complexes because they are spatially separated in the solid. Thus, equilibrium statistics can be used. In other words, the rate of the forward reaction is independent of the rate of the backward reaction. Furthermore, because the activated complexes are spatially separated and do not build up in concentration, classical statistics can be used to describe the activated state.

In equilibrium, according to Boltzmann statistics, the fraction of particles at

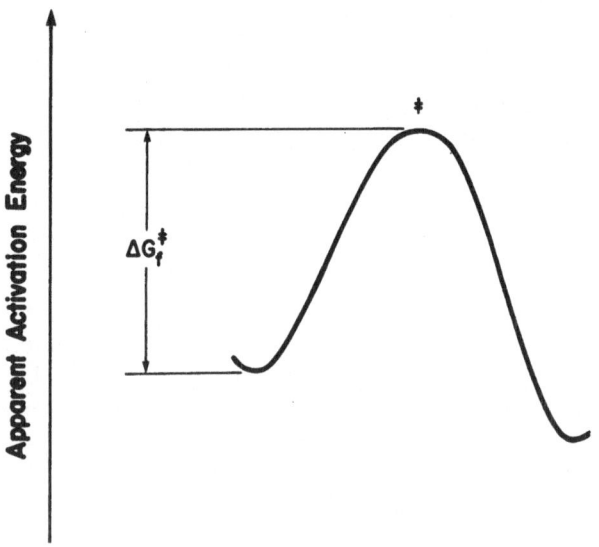

Reaction Coordinate

Figure B.4. Schematic representation of the potential-energy change along the reaction coordinate.

the ith energy level with respect to the total number of particles N_t is

$$\frac{N_i}{N_t} = \frac{\exp(-\Delta E_i/kT)}{\sum_j \exp(-\Delta E_j/kT)} .$$

The introduction of the Boltzmann statistics and further concepts of statistical mechanics leads to the expression of the elementary rate constant [4, 5, 7]

$$\mathscr{k} = \frac{kT}{h} \exp\left(-\frac{\Delta G^+}{kT}\right) .$$

Specific theoretical concepts

The previous relationships were derived for the description of the rate of processes in the absence of an applied force field. The applied force field distorts the potential energy surface in function of the work performed by the system while moving from the equilibrium state to the activated state. In crack growth, as generally for transport processes in condensed phases, there is a relative displacement of ions, atoms, or molecules. Crack growth, and transport processes, correspond to the motion of a point representing the system along the reaction coordinate. The external constraint drives the system. If the work resulting from these constraints is W, then the barrier height changes by the corresponding amount, and the rate constants are

$$\mathscr{k}_f = \frac{kT}{h} \exp\left(-\frac{\Delta G_f^+ - W_f}{kT}\right)$$

$$\mathscr{k}_b = \frac{kT}{h} \exp\left(\frac{\Delta G_b^+ + W_b}{kT}\right).$$

That is, the height of the energy barrier is decreased by W_f for systems moving in the forward direction and increased by W_b in the backward direction. The work W done by the external constraints is partly dissipated as heat and partly converted into the new configuration energy. In crack growth, part of the work expended appears as the surface energy of the increased crack size.

The effect of pressure on the rate constant can be expressed as

$$\mathscr{k} = \frac{kT}{h} \exp\left[-\frac{\Delta G^+ + \int_{p=1}^{p} (\partial \Delta G^+/\partial p)\, \mathrm{d}p}{kT}\right],$$

where p is the pressure. Therefore,

$$\mathscr{k} = \frac{kT}{h} \exp\left(-\Delta G^+ + \int_{p_1}^{p} \Delta V^+\, \mathrm{d}p\right) = \frac{kT}{h} \exp\left(-\frac{\Delta G^+ + \Delta \overline{V}^+ (p - p_1)}{kT}\right)$$

$$= \frac{kT}{h} \exp\left(-\frac{\Delta G^+ + \Delta \overline{V}^+ p}{kT}\right).$$

Comment on the equilibrium assumption

It is sometimes mentioned that the equilibrium assumption of the absolute rate theory casts doubts on its validity. It is important to note that this assumption introduces only negligible errors in the analysis of crack growth conditions. Only at extremely fast crack velocities does it have an effect on the accuracy of theoretical evaluation; a condition excluded from the content of this book. This can be shown as follows.

It has been assumed in the absolute theory of rate processes that the complexes in the stable configuration are in equilibrium with the activated complexes. This assumption expresses only an approximation of the dynamic conditions and is rigorously valid only when the system is in equilibrium. It is very good, however, for most ordinary processes: several investigations have demonstrated that the equilibrium assumption leads to negligible errors. It was found that the error is 20% when $\Delta G^+/kT = 5$, and when $\Delta G^+/kT > 10$ the difference is negligible. Similar conclusions were reached by Eyring and

Zwolinski [6]; Prigogine *et al.* [10] carried out a study of transport processes in dense media using non-equilibrium statistical mechanics. Their results have confirmed the validity of the equilibrium assumption used in the absolute rate theory.

References

1. L. Pauling, and E. B. Wilson: *Introduction to Quantum Mechanics*, McGraw-Hill (1935).
2. E. Wigner and H. Eyring: *Sci. Mon.* **44** (1937), p. 564.
3. R. H. Fowler and E. A. Guggenheim: *Statistical Thermodynamics*, Cambridge University Press (1939).
4. S. Glasstone, K. J. Laidler and H. Eyring: *The Theory of Rate Processes*, McGraw-Hill (1941).
5. H. Eyring, J. Walter and G. E. Kimball, *Quantum Chemistry*, Wiley (1944).
6. H. Eyring and B. J. Zwolinski: *Rec. Chem. Progr.* **87** (1947).
7. T. L. Hill: *Statistical Mechanics*, McGraw-Hill (1956).
8. H. Eyring, D. Henderson, B. J. Stover and E. M. Eyring: *Statistical Mechanics and Dynamics*, Wiley (1964).
9. R. C. Tolman: *The Principles of Statistical Mechanics*, Oxford University Press (1967).
10. I. Prigogine, G. Nicolis and P. M. Allen: in *Chemical Dynamics*. (J. O. Hirschfelder and D. Henderson, eds.), Willey (1971).
11. A. S. Krausz and H. Eyring: *Deformation Kinetics*, Wiley-Interscience, (1975).

The kinetics of crack growth controlled by consecutive energy barriers

First, the simplest, the two-barrier consecutive system kinetics will be discussed, followed by the general case of n barriers.

To provide understanding of the kinetics of the two-energy-barrier system the discussion is formulated in terms of the Region I and Region II concepts of stress corrosion cracking (SCC), presented in Chapter 2. The physical processes of corrosion-bond-breaking and transport steps are considered, for the present purpose, to form a consecutive system of two energy barriers. The rate of activations over each of the barriers is identified by the appropriate elementary rate constant k [1]. The rates of the two processes are identified as

- bond breaking (which may be chemically enhanced), k_{bI}, and healing, k_{hI};
- transport (diffusion for instance), k_{dII}, and its reverse, k_{rII}.*

Figure C.1 represents the typical stress corrosion cracking behavior; Figure C.2 shows schematically the two consecutive barriers.

The total time t_t of overcoming a barrier system is often considered to consist of two periods, so that

$$t_t = t_1 + t_2,$$

where t_1 and t_2 are the waiting times in front of barriers 1 and 2, respectively. A unit reaction of SCC then consists of two elementary steps, each described by the elementary rate constants k_1 and k_2, expressed as

$$t_t = \frac{1}{L_1 k_1} + \frac{1}{L_2 k_2}.$$

This expression is valid only if activation over the energy barriers in the backward direction is negligible. In the complete kinetics description of the SCC process, however, the backward activation also has to be taken into consideration.

---

* The subscript 'd' signifies delivery of the degrading species to the crack tip zone, and 'r' stands for the reverse [2, 3, 5—8]. For the interpretation and application of L, see page 49.

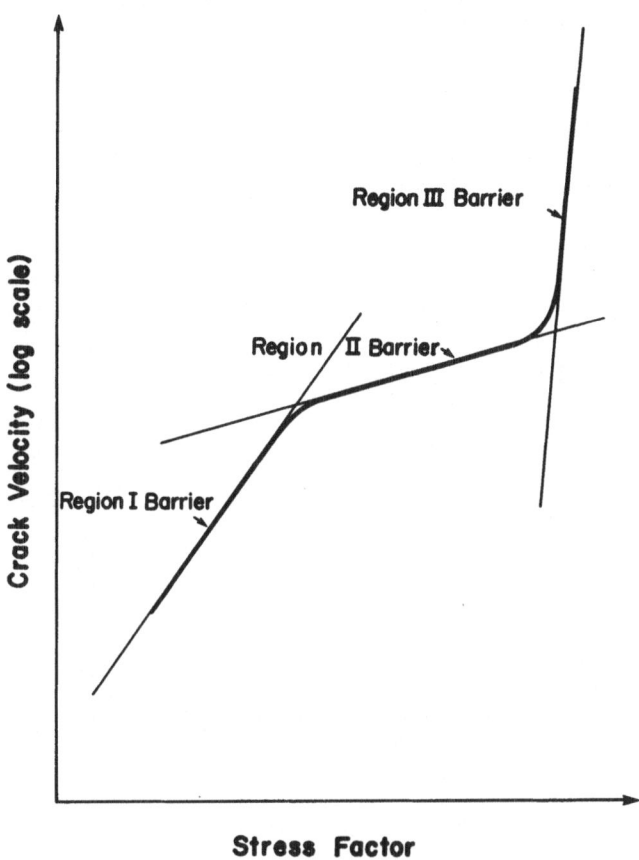

Figure C.1. The combination of the consecutive barriers with the parallel barrier, leading to the observed SCC behavior as a function of the stress intensity.

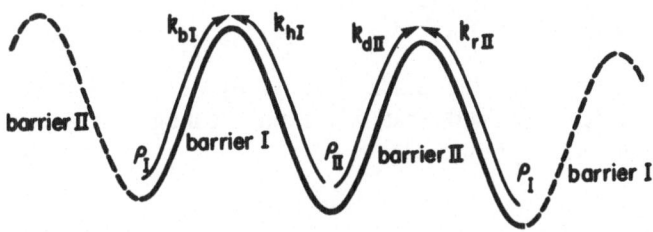

Figure C.2. The schematic representation of the consecutive barrier system of SCC controlled by two barriers. The figure also shows the rate constants with the appropriate notation for the forward and backward activations, as used in the derivation of the kinetics equation.

Consider then ρ_t identical specimens immersed in a homogeneous environment, and subjected to the same constant stress intensity during the full period of crack propagation. In any specimen the crack can be in position I or II, therefore the total number of cracks is [1, 4] (Figure C.2)

$$\rho_t = \rho_I + \rho_{II}. \tag{C.1}$$

The crack velocity

$$v = (L)(\rho_I \, k_{bI} - \rho_{II} \, k_{hI}) \frac{1}{\rho_t} \tag{C.2}$$

is constant for the steady state considered here. In equations (C.1) and (C.2) ρ_t is the total number of cracks; ρ_I and ρ_{II} are the number of cracks with crack tips in positions that correspond to Region I and II states respectively (Figure C.2).

In steady state the net rate of flow over the first barrier is equal to the rate of flow over the second barrier

$$\rho_I \, k_{bI} - \rho_{II} \, k_{hI} = \rho_{II} \, k_{dII} - \rho_I \, k_{rII}. \tag{C.3}$$

Expressing ρ_{II} from equation (C.1) and substituting into equation (C.2) results in

$$v = (L) \frac{1}{\rho_t} [\rho_I(k_{bI} + k_{hI}) - \rho_t \, k_{hI}]. \tag{C.4}$$

The substitution of ρ_{II} into equation (C.3) gives

$$\rho_I \, k_{bI} - \rho_t \, k_{hI} + \rho_I \, k_{hI} = \rho_t \, k_{dII} - \rho_I \, k_{dII} - \rho_I \, k_{rII}$$

and hence

$$\rho_I = \rho_t \frac{k_{dII} + k_{hI}}{k_{bI} + k_{hI} + k_{dII} + k_{rII}}. \tag{C.5}$$

Using equation (C.5), the crack velocity, expressed by equation (C.4), becomes

$$v = (L) \frac{k_{bI} \, k_{dII} - k_{rII} \, k_{hI}}{k_{bI} + k_{hI} + k_{dII} + k_{rII}}. \tag{C.6}$$

Equation (C.6) expresses the crack velocity for the two-consecutive-energy-barrier system. It is noticed that when the healing and reverse rates are negligible, equation (C.6) reduces to

$$v = (L) \frac{1}{\dfrac{1}{k_{bI}} + \dfrac{1}{k_{dII}}} = \frac{1}{\dfrac{1}{L_{bI} \, k_{bI}} + \dfrac{1}{L_{dII} \, k_{dII}}}.$$

in accordance with equation (2.15).

The extension of the two-barrier fracture kinetics description of the consecutive system to that of the *n*-barrier case can be developed by the same technique. The crack velocity expression given in Chapter 2, equation (2.6)

$$v = (L) \frac{1}{\displaystyle\sum_{i=1}^{n} \frac{1}{\ell_i}}$$

can now be written in the complete form: this is left as an exercise [9].

References

1. A. S. Krausz: The deformation and fracture kinetics of stress corrosion cracking, *Int. J. Fract.* **14**, No. 1 (1978), p. 5.
2. B. R. Lawn and T. R. Wilshaw: *Fracture of Brittle Solids*, Cambridge University Press (1975).
3. S. Glasstone, K. J. Laidler and H. Eyring: *The Theory of Rate Processes*, McGraw-Hill (1941).
4. A. S. Krausz and H. Eyring: Chemical kinetics of plastic deformation, *J. Appl. Phys.*, **42** (1971), p. 2382.
5. A. S. Krausz and H. Eyring: *Deformation Kinetics*, Wiley-Interscience (1975).
6. S. W. Benson: *The Foundations of Chemical Kinetics*, McGraw-Hill (1960).
7. S. D. Brown: A multibarrier rate process approach to subcritical crack growth, in *Fracture Mechanics of Ceramics* (R. C. Bradt, D. P. H. Hasselman and F. F. Lange, eds.), Vol. 4, Plenum Press (1978), pp. 597—621.
8. R. P. Wei: in *Hydrogen Effects in Metals* (I. M. Bernstein and A. W. Thompson, eds.), The Metallurgical Society of AIME (1980), pp. 677—689.
9. A. S. Krausz: A deformation kinetics analysis of the stress sensitivity, *Mater. Sci. Eng.* **26** (1976), pp. 65—71.

Supplement to Section 3.5

Using equation (3.5) of Chapter 3, Part 1 [1, 2]

$$\rho_i = \sum_{j=1}^{j=n+1} C_{ij} \exp(-\lambda_j t) + C_i,$$

the first step in the solution of equations (3.36) is

$$\rho_0 = \sum_j C_{0,j} \exp(-\lambda_j t) + C_0$$

$$\rho_{\text{odd}} = \sum_j C_{\text{odd},j} \exp(-\lambda_j t) + C_{\text{odd}} \tag{D.1}$$

$$\rho_{\text{even}} = \sum_j C_{\text{even},j} \exp(-\lambda_j t) + C_{\text{even}}.$$

Each of the j terms, as well as their sums, are solutions and must satisfy the differential equations. For each j, it can be written

$$-C_0 \lambda \exp(-\lambda t) = -C_0 k_c \exp(-\lambda t) - C_0 k_c;$$

$$-C_{\text{odd}} \lambda \exp(-\lambda t) = C_{\text{even}}(k_h + k_c) \exp(-\lambda t) - C_{\text{odd}} k_b \exp(-\lambda t) +$$
$$+ C_0 k_c \exp(-\lambda t) + C_{\text{even}}(k_h + k_c) -$$
$$- C_{\text{odd}} k_b + C_0 k_c;$$

$$-C_{\text{even}} \lambda \exp(-\lambda t) = C_{\text{odd}} k_b \exp(-\lambda t) - C_{\text{even}}(k_c + k_h) \exp(-\lambda t) +$$
$$+ C_{\text{odd}} k_b - C_{\text{even}}(k_h + k_c).$$

Since the time-dependent and the time-independent terms must be satisfied separately, it follows that

$$C_0(\lambda - k_c) = 0;$$
$$C_{odd}(\lambda - k_b) + C_{even}(k_h + k_c) + C_0 k_c = 0; \qquad (D.2)$$
$$C_{odd} k_b + C_{even}(\lambda - k_h - k_c) = 0;$$

and, similarly, for the time-independent terms

$$-C_0 k_c = 0;$$
$$C_{even}(k_h + k_c) - C_{odd} k_b + C_0 k_c = 0; \qquad (D.3)$$
$$C_{odd} k_b - C_{even}(k_h + k_c) = 0.$$

Non-trivial solution of equations (D.2) exists only if the determinant of the coefficients C_0, C_{odd} and C_{even} is zero. The secular equation $D = 0$ is

$$\begin{vmatrix} \lambda - k_c & 0 & 0 \\ k_c & \lambda - k_b & k_h + k_c \\ 0 & k_b & \lambda - k_h - k_c \end{vmatrix} = 0. \qquad (D.4)$$

Expanding the determinant, equation (D.4), leads to

$$(\lambda - k_c)[(\lambda - k_b)(\lambda - k_h - k_c) - k_b(k_c + k_h)] = 0,$$

which can be rewritten as

$$\lambda(\lambda - k_c)[\lambda - (k_b + k_h + k_c)] = 0,$$

and consequently,

$$\lambda_1 = 0; \lambda_2 = k_c; \lambda_3 = k_b + k_h + k_c.$$

The step-by-step substitution of the λ values into equations (D.2) for $j = 1$, $\lambda_1 = 0$ leads to:

$$C_{01} k_c = 0; \frac{C_{odd1}}{C_{even1}} = \frac{k_h + k_c}{k_b} = \alpha;$$

and, if $C_{even1} = a$, then $C_{odd1} = a\alpha$.
For $j = 2$, $\lambda_2 = k_c$:

$$\frac{C_{odd2}}{C_{even2}} = \frac{k_h}{k_b} = \beta;$$
$$\frac{C_{02}}{C_{even2}} = -\frac{k_b + k_h}{k_b} = \gamma;$$

and, if $C_{\text{even2}} = B$, then $C_{\text{odd2}} = B\beta$ and $C_{02} = B\gamma$.
  For $j = 3$, $\lambda_3 = k_b + k_h + k_c$:

$C_{03} = 0$;

$C_{\text{odd3}} = -C_{\text{even3}}$;

and if $C_{\text{even3}} = D$, then $C_{\text{odd3}} = -D$.
  The integration constants C_{ij} are expressed from equations (D.3):

$C_0 = 0$;

$$\frac{C_{\text{odd}}}{C_{\text{even}}} = \frac{k_h + k_b}{k_b} = \alpha;$$

and if $C_{\text{even}} = A$, then $C_{\text{odd}} = A\alpha$.
  The coefficients and the integration constants can be summarized:

j	C_0	C_{even}	C_{odd}
1	0	a	$a\alpha$
2	$B\gamma$	B	$B\beta$
3	0	D	$-D$
C_0	0	—	—
C_{odd}	—	$A\alpha$	—
C_{even}	—	—	A

Substituting these into equation (D.1) gives

$$\rho_0 = \sum_{j=1}^{3} C_{0j} \exp(-\lambda_j t) + C_0 = B\gamma \exp(-k_c t);$$

$$\text{(D.5)}$$

$$\rho_{\text{odd}} = \alpha(a + A) + B\beta \exp(-\lambda_2 t) - D \exp(-\lambda_3 t);$$

$$\rho_{\text{even}} = a + A + B \exp(-\lambda_2 t) + D \exp(-\lambda_3 t).$$

Initially, at $t = 0$ all the cracks are in the first valley, that is,

$$\rho_0 = \rho_t = B\gamma$$

and therefore,

$$B = \frac{\rho_t}{\gamma}.$$

Substituting B and $(A + a) = \varepsilon$ into equations (D.5) gives

$$\rho_0 = \rho_t \exp(-\lambda_2 t);$$

$$\rho_{\text{odd}} = \alpha\varepsilon + \rho_t \frac{\beta}{\gamma} \exp(-\lambda_2 t) - D \exp(-\lambda_3 t); \qquad \text{(D.6)}$$

$$\rho_{\text{even}} = \varepsilon + \rho_t \frac{1}{\gamma} \exp(-\lambda_2 t) + D \exp(-\lambda_3 t).$$

Since at $t = 0$ $\rho_{odd} = \rho_{even} = 0$, it can be seen readily from

$$\rho_{odd} = 0 = \alpha\varepsilon + \rho_t \frac{\beta}{\gamma} - D$$

and

$$\rho_{even} = 0 = \varepsilon(1 + \alpha) + \rho_t \left(\frac{1}{\gamma} + \frac{\beta}{\gamma} \right)$$

that

$$\varepsilon = -\frac{\rho_t(1 + \beta)}{\gamma(1 + \alpha)};$$

$$D = \frac{\rho_t(\beta - \alpha)}{\gamma(1 + \alpha)}.$$

Finally, the substitution of ε and D into equations (D.6) provides the solution of equations (3.36):

$$\frac{\rho_0}{\rho_t} = \exp(-\lambda_2 t)$$

$$\frac{\rho_{odd}}{\rho_t} = F \exp(-\lambda_2 t) + G \exp(-\lambda_3 t) + H \tag{D.7}$$

$$\frac{\rho_{even}}{\rho_t} = I \exp(-\lambda_2 t) - G \exp(-\lambda_3 t) + M,$$

where

$$F = -\frac{k_h}{k_b + k_h}; \qquad G = -\frac{k_b k_c}{(k_b + k_h)(k_b + k_h + k_c)};$$

$$H = \frac{k_h + k_c}{k_b + k_h + k_c}; \qquad I = -\frac{k_b}{k_b + k_h};$$

$$M = \frac{k_b}{k_b + k_h + k_c}.$$

References

1. A. S. Krausz and K. Krausz: The theory of non-steady state fracture kinetics analysis; Part I: General theory of crack propagation, *Eng. Fract. Mech.* **13** (1980), pp. 751–758.
2. A. S. Krausz, J. Mshana and K. Krausz: The theory of non-steady state fracture kinetics analysis; Part II: Non-steady state crack propagation in stress corrosion cracking, *Eng. Fract. Mech.* **13** (1980), pp. 759–766.

Author index

Subject index